This ... to be ... ur ... ...

MICROWAVE PASSIVE DIRECTION FINDING

MICROWAVE PASSIVE DIRECTION FINDING

STEPHEN E. LIPSKY

Senior Vice President
American Electronic Laboratories, Inc.

A Wiley-Interscience Publication
JOHN WILEY & SONS
New   York   •   Chichester   •   Brisbane   •   Toronto   •   Singapore

Library of Congress Cataloging in Publication Data:

Lipsky, Stephen E.
  Microwave passive direction finding.

  "A Wiley-Interscience publication."
  Includes bibliographies and index.
  1. Radio direction finders.   2. Microwave devices.
  I. Title.

  TK6565.D5L57 1987   621.3841′91   86-33995
  ISBN 0-471-83454-8

Printed in the United States of America

10 9 8 7 6 5 4 3 2 1

To two women:

> *My mother, who always said I could*
> *My wife, Hyla, who knew I should*

Foreword

At long last a very professional, complete, and carefully unified consolidation of the latest techniques, both systems and components, is available on the subject of microwave passive direction finding. This volume brings together, for the first time, the latest work done in microwave direction finding by individuals and groups dispersed worldwide. This field is in a period of dynamic innovative growth, spurred by the growing availability of higher-speed lower-cost digital processing circuitry. This book is a welcome and needed assistance to direction finder designers.

LEON RIEBMAN, PH.D.

Preface

This book compiles the many methods of microwave passive direction finding into a single technology, identifiable as such. I have attempted to present the results of the many systems designs and concepts that have evolved early radar direction finding (DF) into a specialized science and state the unique sets of rules and principles that codify microwave passive DF technology.

My method of presentation is tutorial and leads the reader through DF theory in an understandable manner, utilizing mathematics where necessary to understand the concepts and to provide direct design answers. I have found from my experience in writing papers and giving presentations, that most engineers want to be able to reach definitive conclusions based on clear paths of reasoning. To accomplish this, I have described DF technology in theory, by comparison, mathematical analysis, and reference. Where complex questions are to be answered, I have given practical examples. Where many competing factors must be considered, as in the case of the sensitivity of a system for a given accuracy and false-alarm rate, I have made extensive use of computer-developed graphs to simplify the design process. To reinforce this approach, numerous block diagrams, alternate design methods, and illustrations are used.

This volume has been oriented to both the student and practicing microwave engineer. It is presupposed that the reader is familiar with complex variables, some statistics, fields and wave theory, and has a tutorial understanding of radar and general microwave methods. Based on this background, the concept of direction finding is introduced in Chapter 1 as an outgrowth of radar and high-frequency direction finding methods prior to and after World War II. Usage and requirements are stated to show the reasons for the separation of DF technology from its radar origins. Chapter 2 traces the development of monopulse receiver lobing as a natural solution to problems in scanning radar and associated direction finding methods. The postulates and class definitions of monopulse are defined with examples of processing techniques. DF by receiver antenna pattern comparison, introduced in this chapter, is extensively amplified in Chapter 3 by analysis of the many types of antenna elements that achieve required DF phase and amplitude characteristics. Much attention is directed to spiral antenna technology, including multimode operation and extension into the millimeter frequency range. Horn, reflector, and mode-feed developed pattern antennas, such as the Honey–Jones, are presented in various configurations.

Chapter 4 blocks out practical implementations of receivers and processors capable of extracting the DF information from the received signals. Block diagrams of radar warning and ELINT DF systems are used to give the reader a foundation of practical technology. Here may be found descriptions of the unusual: subcommutation, supercommutation, monochannel phase-encoding, and the more commonly known monopulse techniques. Various configurations of "frontends" and receivers used for RWR and ELINT receivers are explained with diagrams of amplitude, phase, and sum-and-difference processors. The three classes of monopulse processors are described with their broadband variants, the detector-log-video-amplifier and the wide bandwidth phase discriminator (correlator). Chapters 5 and 6 cover array technology as applied to receivers. Parallel beam arrays, such as the Butler planar, moded circular, and switched type, are discussed, with examples given of each. Rotman-Turner and R-KR lens-fed arrays are presented to illustrate parallel beam time summation processes that are in common use. Movable reflector, beam, and switched multielement beam-scanning arrays are also described. Interferometer DF systems, which are becoming most important now due to their capability for higher DF accuracy, are detailed in Chapter 5, with practical design equations.

Analytical approaches to the design and analysis of the various receiver types used in passive direction finding are presented in Chapters 7–9. They are intended to define the mathematical aspects of receiver technology in an understandable yet detailed manner that will reward the reader with a comprehensive grasp of some obtuse concepts. Chapter 7 develops methods for signal detection, describing the gain and noise-limited crystal video receiver, the superheterodyne receiver, and downconverter variations of both. Signal-to-noise relationships for pre- and post-detection are derived. Curves permitting determination of signal-to-noise output for signal-to-noise input for various laws of detection are drawn, permitting determination of the sensitivity of a given design. This is done not just for the tangential sensitivity case but for the entire dynamic range of detection using derivations not readily found in the literature. Chapter 8 concentrates on probability-of-detection and false-alarm rates for both the well-established envelope (superheterodyne) and wide-band quadratic (crystal video) receivers. Here again graphical methods are used specifically to relate to the problems of searching for a signal of unknown parameters with a receiver that has characteristics that may only approximate those needed for detection of the intercept's RF and pulse characteristics. Chapter 9, devoted to determining the accuracy of a DF system for amplitude and phase monopulse configurations, is an outgrowth of many articles and papers on this subject. Error budget and channel balance considerations are derived and presented in mathematical and graphical form, with examples of the method of calculation. The effects of noise on accuracy are derived in readily understandable terms, permitting determination of actual operating characteristics of system designs. Practical installation questions are addressed, and recommendations for achieving optimum results are discussed.

Signal analog and digital processing and display methods are examined in Chapter 10, starting with a basic description of three types of logarithmic amplifier designs for real- and non-real-time signal processing, followed by a description of signal processing and concluding with displays and display configurations. The methodology of pulse-on-pulse operation to reduce receiver "shadow" or dead time is described with circuit configurations of practical designs. Wide bandwidth amplitude and phase measurement methods are also analyzed. Complete reference to source material is included throughout the book for further study and investigation.

I have tried to convey my understanding of microwave passive direction finding in a readable form that will encourage technical curiosity and provide the satisfaction that comes with a positive understanding of why this technology is an art.

STEPHEN E. LIPSKY

Rydal, Pennsylvania
May 1987

Acknowledgments

In any book of this type, it is hard to claim originality since most of the concepts and ideas are the work of many teams of engineers working at many different companies. I therefore wish to thank my associates at American Electronic Laboratories, Inc., the General Instrument Corporation, Polarad Electronics, and the Loral Corporation for the opportunity to learn the technology. I am especially appreciative of the support of Dr. Leon Riebman, the Chairman of the Board, and Mark Ronald, president of AEL Industries, Inc., for their faith in and support of this endeavor. I also wish to thank my other AEL associates, in particular, Dr. Baruch Even-Or for his essential assistance and help in Chapters 7, 8, and 9, especially with regard to the derivations and computer plots, and Walter Bohlman, John Bail, and Bob Kopski for their review and suggestions. Additional thanks go to Ernie Buono, Joe Giusti, and the AEL ILS Division for their development of the illustrations, photos, and layouts.

I would like to acknowledge the assistance of Richard Stroh and Stig Rehnmark of Anaren, George Monser of the Raytheon Company, Ron Hirsch of RHG, and Richard Hollis of Watkins Johnson. I appreciate the help and guidance of other respected associates such as Jim Adams of the General Instrument Corporation and Lloyd Robinson at the Stanford Research Institute. I am indebted to Amos Shaham of Elisra, Bene Baraq, Israel, for his descriptions, and to Dr. Donald Linden of the Dalmo Victor Division of the Singer Corporation. The release of technical data, descriptive material photos, and illustrations by the above and many other contributors have made this book possible. I also wish to thank Mr. George Telecki, my editor at Wiley, for his faith, encouragement, and patience.

Last but far from least, I wish to thank Judith Butterfield for her help, unending patience, and diligence in the preparation of this manuscript. In any endeavor of this sort no one is alone. I thank my wife, Hyla, for her encouragement and forbearance, without which this book would have neither been started nor completed.

STEPHEN E. LIPSKY

Contents

MICROWAVE PASSIVE DIRECTION FINDING

Evolution and Uses of Passive Direction Finding

In recent years, microwave passive direction finding (DF) has emerged as a distinct technology apart from HF/VHF and radar associated DF methods. This has occurred as the result of many factors, the most important being recognition of electronic warfare as an art. In the post-World War II period, radar-controlled weapon systems have undergone extensive development aided, in no small part, by the enormous worldwide investments and subsequent achievements in avionics, semiconductor, and radar technologies. In the initial stages of this expansion, the use of direction finding, both active and passive, was part of the more general task of gathering electronic battlefield intelligence (ELINT). At first, the recognition of the number and types of hostile emitters, in a field of many signals, was accomplished by low probability-of-intercept systems, typically using rotating antennas, narrow bandwidth scanning receivers, and real-time pulse processing and display techniques. This methodology was consistent with speed of the available data-handling processes of that time and the human operator's ability to make use of the information.

More recently, with the advent of sophisticated high-performance antiaircraft and antiship missile systems, electronic warfare has been tasked with the problem of recognizing not just the threat but rather the state it is in: acquisition, launch, and control or terminal guidance. The need for early warning and protection for ships, high-speed aircraft, and ground forces has led to the development of complex electronic warfare protection suites and a class of equipment known as radar warning receivers (RWR), both of which make extensive use of microwave passive DF methods. Reaction time has also become a forcing function since there may be only minutes or seconds between detection and destruction, often requiring that a decision be made automatically without a human operator. To make matters more complex, the proliferation of equipment sold, captured, and exchanged between hostile countries, formally friendly to each other, has all but destroyed recognition of the radar equipment of a friend or a foe on a "country developed" type basis.

1

The answer to this current situation has been the development of real-time electronic warfare techniques utilizing high probability-of-intercept computer-controlled systems to sort out the environment on a pulse-by-pulse basis over wide-frequency bandwidths and full 360 degree azimuthal fields-of-view. The signal density and quantity of data obtained by high-probability intercept receivers had, in the past, completely outstripped the then available processing capabilities. Fortunately, the same technology that has exacerbated the threat determination problem has improved digital capacity by providing more storage in smaller volumes and at faster throughput rates, making recognition and sorting of signals on a pulse-by-pulse basis now practical. In the microwave radio frequency (RF) area, perfection of wide bandwidth instantaneous frequency measurement has added an additional impetus to the need for instantaneous intercept DF data since both frequency and DF information can be obtained on a single pulse basis and used as presorting tools, using one to direct the other to reduce system processor workloads.

Improvements in antennas and the availability of matched microwave solid-state RF amplifiers now permit greater ranges of detection and higher accuracies to be obtained over wider instantaneous bandwidths and at higher sensitivities. These advances have made microwave passive DF technology self-adaptive as compared to active HF and microwave radar systems that were greatly aided by a priori knowledge of frequency, anticipated direction, or pulse characteristics of a known illuminating transmitted signal. Passive direction finding has emerged as a stand-alone technology completely adaptive to complex unknown environments and fully capable of operating with the multitudinous types of signals found in modern electronic warfare.

1.1 EVOLUTION

This book has been written in recognition of the development of microwave passive direction finding as a basic technology permitting one to develop a synergistic view of the many concepts to permit the selection of the technique that best fits the problem to be solved. The material to be covered has been selected from a very wide range of published, known, and new techniques that meet an essential rule—that of practical application. There are many DF methods based upon theoretical projections of performance, idealized antenna characteristics, computer assets, and so on, that remained submerged in the world of useful technology. In no other field perhaps have there been as many false starts, aborted designs, and abandoned systems due perhaps to the reactive method by which electronic warfare equipment has evolved. The appearance of a threat is countered by a threat recognition system until the next threat is discovered. A major factor in the process is the constant need to improve the probability of receiving few pulses on short signal bursts since radiation security dictates that radar and target determination systems minimize on-the-air time. Passive detection systems have therefore been designed to meet this challenge. The high density of signals has mandated computer sorting and control as a necessity.

Passive direction finding has also emerged for weapon targeting. As the accuracy of DF systems improves (2 degrees root mean square (RMS) accuracy seems to be the magic accuracy number), missiles and aircraft can use the passive DF data to steer along the line-of-bearing to intercept targets, using terminal guidance methods for high accuracy at the last stage. The technology of microwave direction finding is undergoing constant pressure to increase DF accuracy for this purpose.

This book will present the development of passive DF from its early origins to present-day methods to include monopulse, beam-formed arrays, interferometers, and rotating antenna systems. Operation of important subsystems that affect overall performance will be analyzed, with mathematical models presented for key factors. Determination of signal-to-noise ratios, detector operation, monopulse ratio normalization, DF accuracy, and intercept probability will be detailed to give a synergistic understanding of the system aspects of the solution.

1.2 HF DF ORIGINS

Passive direction finding was the first type of target location system based upon radio techniques. The major advances in this art were derived from the needs of both world wars. It is interesting to note that much of the technique development and patent protection was filed in the 1920–1940 time period, when direction finding was concentrated in the "shortwave" or 0.5–30 MHz frequency range. Initial concepts recognized that the nonisotropic or directional pattern characteristic response of one or more antennas could be used to determine the direction-of-arrival of a signal by noting the pointing position of the receiver's "imperfect" antenna when the received amplitude peaked or nulled as the antenna was rotated. Emitter location by triangulation, the intersection of three or more bearing lines, required more sophisticated directional antennas and new methods to achieve improved bearing accuracy. Fortunately, most signals remained on the air long enough to permit multiple bearing measurements to be made, either by a single receiver moved to different locations or by several geographically dispersed receivers tuned to the same frequency and, it was hoped, to the same signal. Continuing needs for better accuracy soon led to antennas with better performance, that is, more signal change per degree of rotation. Null type systems, such as the loop antenna and multiple element arrays, permitted remarkable accuracies to be achieved, in some cases to the extent of the limit of errors due to summation of the ground and sky wave propagation paths. This latter limitation in the HF range turned out to be a major problem. The sky wave or "skip" component arrived randomly polarized causing serious null precession or shift when added to the well-behaved vertically polarized ground wave signal.

Adcock (1), working in England in 1919, patented a narrow-aperture (shortened length) array of four orthogonally located elements, which, when summed and compared, could give the effect of two crossed figure-eight patterns as shown in Figure 1-1. Comparisons of the signals provided DF field patterns that had a 180 degree ambiguity, which was resolved by comparison to the sum of all of the antenna patterns, which was omnidirectional. Sometimes the signal from another

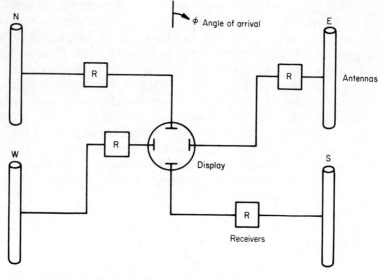

a) System of four monopoles

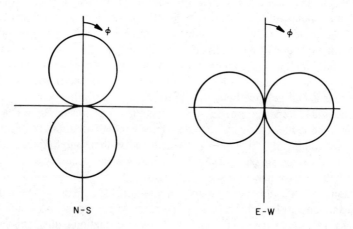

b) Display patterns of a)

Figure 1-1. Adcock antenna and resulting patterns. (*a*) System of four monopoles. (*b*) Display of patterns of (*a*).

element, called a sense antenna, would be used to provide a pattern that would give a cardioid, or single-null response. This technique was also used with loop antennas for the same ambiguity resolution purpose. The German Wullenweber array and others improved matters by using longer medium- and wide-aperture arrays of as many as 36 elements dispersed around a large diameter circle. A set of one third of the antenna elements was passed through a capacitively coupled rotating joint through an RF delay line and summed in two groups of four elements each, as shown in Figure 1-2. The two groups were subtracted with the resultant being a null (Fig. 1-2*b*) that physically rotated at the mechanical rotational rate of the mechanism, known as a goniometer. When connected to a suitable receiver, DF measurements could be made to fractional degree accuracy, as limited above (see Ref. 2).

During the period from World War II to the present, the Adcock and similar antenna systems were improved. It is worth noting, for future reference to microwave designs, that the improvements to these systems moved in the direction of developing

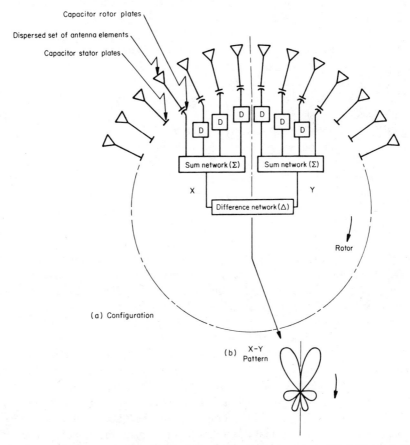

Figure 1-2. A goniometer rotating capacitor transformer used to develop a rotating null for DF determination.

both the difference and the sum patterns of grouped elements and a means of comparing the two simultaneously to remove common signal amplitude variations, a technique that we shall see is similar to radar "monopulse" methods.

1.3 RADAR DF ORIGINS

Direction finding in the microwave frequency range, defined loosely here as 500–100,000 MHz, derived from two sources: the HF direction finding systems described above and the invention of radar, the technique of radio ranging and direction finding. Transition of operating techniques from the HF to the microwave region was readily accomplished since greater aperture or wave interception area for the antennas could be obtained because microwave wavelengths were smaller at the higher operating microwave frequencies. Propagation of microwave energy was by line-of-sight, not by reflection from the ionosphere due to the higher frequencies involved, fortunately eliminating multipath "skip" problems.

Radar development was rapidly advanced during the late 1930s in Germany, England, and the United States in anticipation of World War II. Stories about the use of radar in the battle of Britain are legend as are some of the radar countermeasures developed as a necessity. This book will not attempt to document each step but will rather identify the key factors leading to the present concept of passive direction finding as applied to electronic warfare.

Early radars were generally monostatic (collocated transmitter and receiver) using the same antenna system to transmit and receive a series of pulsed signals. Short pulse signals allowed high peak powers to be transmitted for maximum range. The interval between the pulses (the pulse repetition interval PRI) was chosen to provide unambiguous range returns. The time difference between the transmitted signal and the receipt of the target echo was measured to extract range information. To obtain angle information, the antenna was rotated through a known angle to cover a desired azimuthal sector, and the pointing direction at which the radar return was received was noted. Tactical needs to increase range measurement led to the development of higher power transmitters, high gain, more efficient antennas, and better angle-of-arrival techniques. As components improved it was possible to make advances in peak power generation by shortening the transmitted pulse widths. DF resolution was improved by the use of higher frequency narrow beamwidth antennas. This required better scanning methods and the development of scanning or sequential lobing antenna feeds to track or follow a specific target for gun aiming purposes or to present a polar map of all signals (plan position indication (PPI)).

New requirements called for the radar to view a sector in azimuth broadly while simultaneously moving or scanning the antenna feed slightly in a periodic fashion to "paint" across, or modulate, the target (called nutating), illuminating it at a known rate or scan. If the radar return scan illumination was equal to the transmitter scan illumination as determined by comparing the phase and amplitude of the scan modulation of the return to that of the transmitted signal, the radar was deemed to be on "boresight." If the return scan modulations were unequal, an error signal was developed and negatively fed back to the antenna servo, or positioning system, to effect a correction. By scanning in either a circular or conical mode, it was possible to develop both azimuth and elevation correction signals, allowing for full tracking or "track-while-scan" operation. Sequential transmitter lobing accomplished the same effect as nutation by switching between two squinted or physically displaced transmit beams at a prescribed rate. In both of these cases the scan modulation was contained in the transmitted beam, and the symmetry and amplitude of the illuminating beam could be detected at the *target* to indicate a "track" or "lock-on" condition. The symmetry of the received scan at the radar was a measure of nearness to boresight of the radar. A reduction of the amplitude of the scan would induce the use of more radar system gain to "tighten" or better lock the radar's antenna aiming servo.

Early in the development of radar certain problems of resolution error became evident. In a sequential or scanning radar the received signal consisted of a series of pulses, differing in amplitude by the scan rate. If there was any pulse-to-pulse variation not due to the intended scan, there were errors. This happened as a result of target movement or scintillation. Angular jitter due to multiple reflections in the path of the illuminating beam caused substantial variations. While range gating, which is the process of turning the radar receiver on only over the time period corresponding to the anticipated range of the expected return, helped this problem, another method was needed. A new DF measurement technique was developed to form and detect the received beam by simultaneously receiving it in two separated but collocated receiver antenna lobes. The pulse-by-pulse ratio of the received pulses in each beam would then contain all of the DF information, independently of any pulse amplitude variation. The approach was aptly named monopulse. There is an important difference in this technique in comparison to scanning pulse radars: With monopulse radars the transmitter emits a constant amplitude signal, with DF information being extracted from the ratio of the pulse amplitudes (or phases) of two or more receiver beams that are angularly (or phase) offset in a known manner while being physically scanned. The DF lobing or scanning is not contained in the transmit beam as before. Since this technique, known as passive lobing, can be done on a single pulse, DF measurements can be made on target-reflected pulses containing intrapulse variations, since it is the *ratio* of the two receiver beams that holds the DF data. This ratio, as will be shown, normalizes the intercept return removing all intrapulse or common-mode variation, resulting in a very basic improvement.

The art of passive DF determination was an outgrowth of the development of monopulse radar. The transmit beam was only used to illuminate the target; all DF measurements were made by the receiver. Since the receiver monopulse antenna or array was corotated and pointed with the transmitter antenna (called monostatic), the radar system had an advance idea of where the target would be by knowing the aiming direction of the antenna system. As receiver sensitivities improved, it often became convenient to separate the transmit antenna from the monopulse receiver system and not collocate the two, leading to the development of the separated (or bistatic) radar where the transmitter "illuminated" the target from one location while the receiver detected the angle of arrival of the return from another. By further improving the receiver sensitivities and associated antenna system, signals normally transmitted by the target, such as its own radar, could be used as the returns, eliminating the need for the "illumination" transmitter altogether. Under these conditions, no emissions took place at the receive point, the DF technique was termed "passive," and the art of passive direction finding was born. It is this part of the overall radar technology that will be examined in detail in this book.

1.4 USES OF PASSIVE DF

Some of the more important uses of microwave passive direction finding relate to modern electronic warfare applications. It is relevant to discuss some of the more significant ones.

Early Warning Threat Detection. Early warning signal or threat detection is perhaps the chief use of passive DF. With high-speed processing it is possible to detect and identify a threat in a dense environment by determining the status of the threat amplitude, scan, pulse width, and period modulation. From this information it is possible to recognize the presence of an imminent attack and permit countermeasures to be taken. One of the first uses of passive direction finding was to locate the positions of ships and submarines by shore stations and by the submarines themselves to protect against antisubmarine patrols. In World War II this was an important activity. Allied DF methods were well advanced due in part to the rescue of an IT&T DF research group from France just before the fall of Paris (3).

German submarines, in the latter part of the war, were equipped with microwave receivers and crystal detectors, operating at L-band radar frequencies, to give early warning of British antisubmarine radars. Most of these systems, primative by today's standards, were hand directed and limited by the sensitivities obtainable. Crystal diode video receivers, with sensitivities of only −45 dBm (decibels below one milliwatt of received power in 50 ohms) sensitivity and some antenna gain, gave suitable warning in many cases, however, since a radar signal travels to the target and back, suffering double attenuation in addition to the target reflection or return loss. This made the radar signal considerably stronger at the target than at the radar, requiring less target receiver sensitivity. As a result, British RAF L-band shore radars used to detect submarines were detected themselves by the submarines at ranges beyond that of the radar, giving them time to escape. It was only when the British moved their radars to S-band and the Germans were unable to detect them were the tables turned (4). The principle still prevails for radar warning receivers where simple crystal video detectors are often used to provide detection at approximately twice the radar's detection range.

Early line-of-sight aircraft DF systems were relatively successful since the higher altitude of a plane could give warning out to the radar horizon if receiver sensitivities were sufficient. Much of this early effort has been well documented in Ref. 5. The problem of detecting the scanning victim radar with a rotating DF receive antenna was severe since the detection probabilities resulting from two rotations—that of the receiver and that of the radar (numbers less than 1)—multiply. The need for sensitivities high enough to receive the backlobes of the radar antennas, which were about 30 dB below the beam peak, was finally achieved, partially removing the effects of the radar antenna rotation. This increased the dynamic range of receiver operation, necessitating the development of compressive or "logarithmic" amplifiers, which became a key factor in the development of monopulse technology.

Targeting, Homing, and Jamming. In modern warfare, missiles are usually launched in the general direction of the target by a radar system that acquires the target, feeds initial coordinates to the missile prior to launch, and then continues to transmit guidance signals to the missile along its way. The missile system acquisition and guidance radar reveals itself, however, by the need to radiate energy. A passive DF system can be used as an "antiradiation" technique for DF steering to a target radar. For targets that shut off their radars to prevent detection, passive DF can bring the missile to a range near the target before it shuts down, after which a self-contained terminal guidance missile-borne radar can be activated to guide the missile to hit the target before countermeasure action can be taken. The relative physical small size of a missile necessitates the use of correspondingly small size on-board antennas operating at higher frequencies to reduce wavelength-associated dimensions. These concepts have prompted the development of small high-accuracy passive DF systems.

Repeater type jammers are concerned with generating a deceptive return signal at high power on the true target return, usually within the backporch of the illuminating pulse. This is done to modulate the echo scan falsely, causing loss of lock at the radar. To accomplish this, it is necessary to build retrodirective DF systems with high angular accuracy for both receiving and transmitting antennas, since narrower antenna beamwidths used to give better antenna gain require that the antenna be pointed precisely. As a result, more accurate DF systems, depending upon arraying or combining of many antennas, are used to obtain larger apertures for higher angular resolution, rather than depending on the pattern accuracy of individual DF antennas. This new technology makes use of lens-fed or switched-phaseshift arrays to form and steer narrow beams and is a factor in the development of DF technology apart from radar.

Electronic Intelligence. Electronic Intelligence (ELINT) is an important use of passive direction finding since, because no energy is radiated, passive DF can be used discretely to calibrate the electromagnetic spectrum continuously. In wartime ELINT determines the electronic order of battle (EOB); in peacetime it is the daily activity of updating or mapping the environment for the purposes of detecting new signals and changes of deployment of threats or forces, as well as providing a general measure of radio and radar traffic. A radar transmitter, for example, emits a signal that identifies itself in many ways. RF frequency, pulse width, pulse interval, and signal grouping provides real-time descriptors that can identify the type of radar in use. Scan or lack of it, scan frequency, and received power add additional information. Indeed, in electronic intelligence applications, it is possible to codify types, classes, and names of radars based upon observation of these parameters and to use this information as a "telephone directory" or "look-up" table when identification is needed. Radar is active; it reveals itself and in so doing provides a means to be countered.

The relative lack of published theory and information about passive direction finding derives from its counterpurpose to radar and other signals. In general, passive direction finding adds the missing dimensions to defensive electronic warfare, determination of the angle-of-arrival, and identification and subsequent location of radiating signals without revealing DF receiver presence or operating technique. Passive DF is useful in quiescent peacetime situations to calibrate a signal environment by determining location of known signals. It is essential in wartime to provide recognition of the signals and defense against radar and other RF directed threats either by warning of their presence or by steering a jamming signal, missile, or dispensing countermeasures. With these objectives, passive direction finding always strives for maximum accuracy in the least possible time as a primary consideration. Performance tradeoffs must naturally be made in both cost and installation; however, the methods and technology of passive direction finding provide a challenging contrast to the well-known and accepted technology of radar.

Data Reduction. Despite the improvements cited earlier, digital data rates are still the limiting factor to better DF processing. There are simply too many signals to process at one time. The "down" or "shadow" time of an electronic support measure (ESM) system, can be improved by waning or thinning the environment by true angle-of-arrival using frequency as presorting descriptors to limit data flow. Although this approach can make a system blind at times, the actual refresh or update rate of an operator-assisted ESM system can be dramatically improved. To obtain these desired results, it becomes important to improve accuracy. Consider a received radar intercept: As the radar scans, each pulse varies in signal strength from its predecessor throughout the transmitter's antenna beamwidth. This causes each pulse to be received at a different signal-to-noise ratio by the passive DF system. In simultaneous amplitude comparison DF receivers, the DF determination rules can change unless signal strength is well above a preset threshold. If care is not taken, a single signal can appear as a multitude of signals differing in many characteristics, as the signal-to-noise ratio varies causing errors, and making signal separation extremely difficult. The solution for this is averaging or adding processing gain to compensate; however, this takes time. The problem may be tolerable for ELINT but intolerable for threat warning.

The availability of instantaneous frequency information can also be used to sort signals quite accurately and in some instances can correct for any known bearing versus frequency anomalies that are repeatable. This latter feature is especially useful for aircraft that can be calibrated under anechoic conditions. Digital look-up tables can also be used to extend the range of frequency ambiguous systems, such as interferometers, by recognizing that null shifts, for example, will occur at certain frequencies that are half-wavelength multiples.

1.5 SUMMARY AND GUIDE TO THE BOOK

This first chapter has introduced the concept of microwave passive direction finding as a science apart from radar receivers, the purpose being to permit a synergistic examination of the theory, antennas, and techniques that constitute the technology. The roots of modern passive direction finding evolved from early HF/VHF and radar methods. Solutions, found to solve radar problems, have been used to mold passive direction finding into a more sophisticated unified art, which, in combination with digital methods, has made this technique a key factor in present-day electronic warfare systems.

The remainder of the book has been divided topically. Chapter 2 develops DF receiver theory and the concepts of monopulse. Chapter 3 details antennas for DF applications, describing operation modes, theory of operation, and key details of design. Chapter 4 shows how various receivers can be configured to extract the DF data optimally. The important concepts of receiver antenna arrays are explained in Chapter 5. Interferometer antennas, a special form of arrays, are covered in detail in Chapter 6.

The mathematics of the DF process are presented in Chapter 7, which explains detection theory for wide and narrow bandwidths for the various operational laws of both linear and quadratic detectors. Chapter 8 develops detection probability theory as it specifically applies to the problems of direction finding, with many useful graphical methods and examples. DF accuracy and how to compute it are explained in Chapter 9 for both phase and amplitude monopulse systems. Chapter 10 describes wide bandwidth logarithmic video, RF, and pulse-on-pulse video amplifier methods and shows several human–machine interactive displays for DF presentation. The book concludes with predictions for future systems, outlining some of the newer techniques for millimeter-wave coverage.

REFERENCES

1. Adcock, F., "Improvement in Means for Determining the Direction of a Distant Source of Electromagnetic Radiation," British Patent 1304901919.
2. Gething, P., "Radio Directing Finding and the Resolution of Multicomponent Wave-fields," *Proc. IEE British Electromagnetic Wave*, Series 4. Stevenage, Herts., England: P. Peregrinus Ltd., 1978.
3. "Huff-Duff vs the U Boat," *Electronic Warfare*, May/June 1976, Vol. 8, No. 3, pp. 72, 73.
4. Morse, P. M., and G. E. Kimball, *Methods of Operations Research*, New York: Wiley, 1951.
5. Price, A., *The History of U.S. Electronic Warfare*, Westford, MA: Association of Old Crows, Sept. 1984.

DF Receiver Theory

Microwave passive direction-finding technology has evolved from radar and earlier methods in a subtle way; we still trace this development from its rotary antenna and radar receiver origins to modern monopulse methods developed to provide sophisticated DF accuracy improvements. The evolution took place in answer to limitations in radar angle detection techniques that became evident as radar systems became widely developed. Problems included undesired lobe pickup, spurious responses to signals, return "glint," and limited dynamic range. First, solutions were to construct antennas with highly directional properties (very high front-to-back lobe responses) and multichannel receivers with special characteristics leading to the development of the relatively complex monopulse concept as we know it. Monopulse, however, was not the final result. Many radars designed for long-range surveillance, navigation, weather status, and ship positioning still use a single-channel antenna and receiver for simplicity and low cost. In situations in which well-trained operators are available, environments are light, and low cost is essential, nonmonopulse radars and associated single-channel DF receivers are frequently used. In situations in which high precision and immunity to jamming are of primary importance, monopulse becomes the technique of choice. As a consequence, the technology of passive direction finding must detect and recognize both monopulse and rotating radar techniques.

2.1 EVOLUTION OF ROTATING DF SYSTEMS

What are the problems of a simple single-channel rotating DF antenna system? Consider the DF antenna as it spins continuously, intercepting, displaying, and/or encoding all received signals as shown in a typical antenna beamwidth pattern in Figure 2-1. This particular antenna has a gain of G_p and a 3 dB beamwidth W,

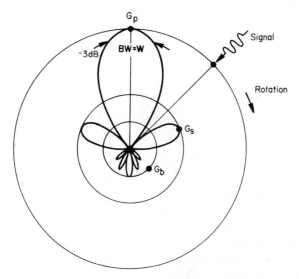

Figure 2-1. Pattern of a typical rotating DF antenna.

which is made as narrow as is consistent with rotation rate and desired accuracy. (Gains of 12 dB beamwidths of 30 degrees and rotation rates from 150 to 1500 rpm are typical.) The first sidelobe response level exhibits a gain G_s, with a maximum backlobe gain of G_b width values of 15 and 30 dB, respectively, as typical. The usable unambiguous dynamic range of this antenna can be seen to be

$$\text{antenna dynamic range} = (G_p - G_s) \quad \text{in dB}$$

This is the case for signals that do not overload the antenna or receiver and that are above the desired threshold of detection. It is also assumed that the signal amplitude variation and pulse parameters are within appropriate limits to permit detection throughout at least one full receiver antenna rotation to assure that the receiver DF beam maximum is displayed.

The DF display viewed on a high-persistence polar cathode ray tube (CRT) shows signal strength of the various received radar pulses versus angle of arrival and will be a pattern of the type shown in Figure 2-2. This depicts the signal unambiguously at its incoming azimuth at the width of the DF antenna's beamwidth, which is filled with pulses at the signal pulse repetition frequency (PRF). The antenna side-lobe level is just visible since the instantaneous dynamic range of the display is approximately that of the antenna main beam to first lobe level.

Let us assume that the signal is increased in strength enough to saturate the main lobe and more fully enter the side-lobe level, as shown in Figure 2-3. The display changes: The mainlobe spreads out, the side and backlobes become more evident, and the screen becomes filled with signal strobes. Determining the exact angle of

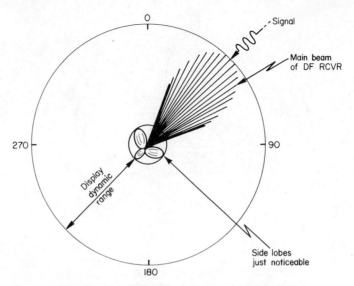

Figure 2-2. A CRT display of a simple direction finder.

arrival of the intercept now becomes more difficult, and there are serious questions about bearing accuracy and ambiguity. If other strong simultaneous signals appear at different bearings, they will be received in much the same way as the first, further complicating the display.

A good operator can position the instantaneous dynamic range of the system on the usable part of the display by adjusting the gain to attempt to identify multiple

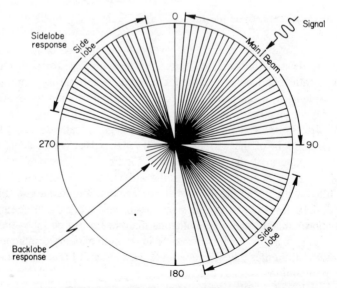

Figure 2-3. Simple DF display when receiving strong (saturating) signals.

intercepts by comparison of their relative amplitudes over at least one DF rotation. The maximum visual dynamic range, however, is the ratio of the maximum strobe-length display radius to minimum discernible center strobelength, which, for a linear system (using reasonable diameter CRTs), is unfortunately not large. In most cases, this prevents association of the strobes from multiple intercepts with their specific signals. If an extremely strong signal enters the antenna's back lobes by exceeding the total dynamic range of the antenna $(G_p - G_b)$, the system is effectively jammed. We have therefore identified two distinct problems common to simple rotating DF antenna/receivers: undesired lobe response and limited dynamic range. Let us now see how the solution to these two problems led to the use of the tools of the monopulse radar art.

2.1.1 Side and Backlobe Inhibition Techniques

A method has been developed to reduce or inhibit the undesirable lobe responses, shown in the single-channel system of Figure 2-1, by the use of two antennas and receivers in a dual-channel system. Here the first antenna is the same rotary DF type as before, but the second is omnidirectional (omniazimuthal), designed to cover the same spatial elevation angle (H plane) having, as a consequence, a lower relative gain. When each antenna is connected to one of a set of receivers that are amplitude matched over the desired frequency range, the relative gains of the two channels can be adjusted to position the omniazimuthal channel gain between the DF main and sidelobe levels, as shown in Figure 2-4, such that

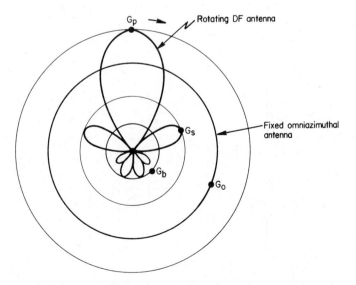

Figure 2-4. A two-channel omniazimuthal sidelobe inhibited DF antenna.

$$G_p > G_o > G_s > G_b$$

Under these conditions, simple logic can be used to permit acceptance of only those signals that meet the above criteria; namely, DF output will be indicated only when the DF signal is greater than the omniazimuthal output, which in turn must be greater than the DF side- and backlobe levels by definition.

What has been described is a well-known skirt-inhibition technique (2) found in many angle measurement systems. Certain additional assumptions have been made: The initial antenna gains maintain their relative relationships or track each other over frequency, axial ratio, fields of view, and polarization; the receivers are amplitude matched and exhibit the same bandwidths and dynamic ranges; and there is no signal present that is of sufficient magnitude to jam the system. In this latter case, the interfering signal would have to saturate both receivers, making their outputs independent of the antenna patterns. Subject to these limitations, the inhibition technique works but, unfortunately, only over a limited dynamic range, as discussed above.

2.1.2   The Logarithmic Amplifier as a Dynamic Range Solution

The need for a solution of the dynamic range problem has led to the application of the logarithmic intermediate frequency (IF) amplifier to passive direction finding systems. Logarithmic IF amplifiers, originally developed for superheterodyne radar receivers, provide an arithmetic voltage output for a geometric RF signal input to compress the input DF dynamic range. Figure 2-5 is a representative characteristic curve showing this relationship. From the curve, it may be seen that for an input

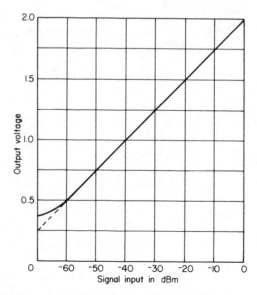

Figure 2-5. Representative logarithmic amplifier transfer characteristic response.

RF variation of 60 dB there is an output voltage variation of only 1.5 V, representing a substantially reduced output range. The amplifier has effectively compressed the dynamic range of the input signal to more manageable proportions, since a 60 dB voltage range would swing from microvolts to volts. (It should be recognized that the logarithmic intermediate frequency amplifier shown here is actually a detector and amplifier combined, since the RF carrier is demodulated in the process.) Although compression reduces the important front-to-backlobe ratio, when used in the DF systems described above, careful control of the gain match and transfer characteristics of logarithmic amplifier pairs can permit the logic requirements we have imposed to be met in practice.

First designs of radar receiver logarithmic amplifiers were fixed frequency tuned intermediate frequency types and were used in superheterodyne radar receivers to provide optimum signal-to-noise ratios for the known pulsewidth of the radar return. Operating frequencies were typically 30, 60, or 100 MHz, with instantaneous video bandwidths of 4–10 MHz optimized for the best signal-to-noise ratio of the transmitted pulsewidth returns. For most wide RF bandwidth DF systems, however, intercepts must be received with pulse parameters that fall within wide unknown limits, not permitting optimum receiver bandwidths to be used, thus necessitating the development of a second type of logarithmic amplifier—the detector logarithmic video amplifier or DLVA. In this approach a wide RF bandwidth square-law diode detector is combined with a wide video bandwidth compressive amplifier to attain known and predictable logarithmic characteristics. The DLVA finds extensive application in multichannel crystal video direction-finding applications, where maximum intercept probability is desired for intercepts with unknown frequency and pulse characteristics. A wide video bandwidth, often greater than a radar receiver would require, is used to assure reception of the narrowest pulses. This is necessary to receive a range of unknown pulse widths. An instantaneous RF bandwidth of greater than 20 GHz for nanosecond video pulse widths is readily achievable over 60 dB dynamic ranges in typical DLVA radar warning receiver applications.

2.1.3 Wide Dynamic Range Skirt Inhibition

The next step in the evolution of direction-finding receivers combined skirt inhibition with the dynamic range solution provided by the logarithmic amplifier. Figure 2-6 is a block diagram of a lobe-inhibited passive DF system using the dual-channel scanning superheterodyne receiver and the same rotating and omnidirectional antennas described previously. Signals received by the rotating DF antenna pass through a single-channel rotary joint where they are mixed with one of the outputs of a local oscillator, made common to both channels to assure that each receives the same signal. The resulting DF intermediate frequency signal is amplified by one of two identical logarithmic IF amplifiers, where it is detected and converted to a video signal, which is fed, through a variable attenuator (A), to the DF gate circuit. Similarly, the omnidirectional (omni) antenna receives the same signal, which is mixed with the second output of the local oscillator and converted to omni video by the action of the second matched logarithmic IF amplifier. The logic circuitry,

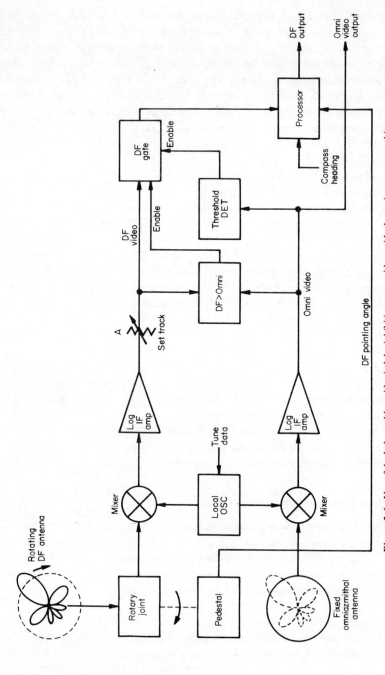

Figure 2-6. Use of dual-channel logarithmic lobe inhibition to provide a wide dynamic range, unambiguous DF output.

consisting of a threshold detector and decision circuitry, allows the DF gate to pass DF video only when the following conditions apply:

1. The omni signal is present at a level that attains the desired system signal-to-noise ratio for a given false alarm rate.
2. The DF signal is greater than the omni.
3. The system is not jammed.

The DF signal is fed to an encoder/processor, which can also accept a compass heading to provide a DF output that can be displayed as either true or relative to the host vehicle or site. The omniazimuthal video is made available for pulse time of arrival, amplitude, and pulse width analysis. The receiver bandwidth is determined by logarithmic IF amplifiers and is chosen to pass pulsewidths expected in the frequency range to be covered. The use of a relatively narrow bandwidth super-heterodyne receiver necessitates a frequency tuning or scanning procedure, which is accomplished by varying the frequency of the voltage-tuned local oscillator by changing its tuning data. When a signal is intercepted, the receiver scan stops to permit measurements to be made. All signals above the established signal-to-noise ratio that also satisfy the DF > omni antenna criteria will be received, assuring proper backlobe inhibition. DF measurement is made by taking one-half of the DF antenna angle throughout which signals have been received, essentially bisecting the symmetrical rotating DF antenna beamwidth. It must be recognized, however, that in the system described here, target scan will add error by distorting the received signal amplitude and hence symmetry of the DF beam.

To ensure the proper gain relationships between the DF and omniazimuthal antenna and to provide lobe inhibition, a variable attenuator (A) is adjusted to place the omniazimuthal gain between the DF main lobe and first sidelobe responses, as shown previously in Figure 2-4. Since this relationship can change with tuning and other factors, an electronically programmable attenuator may be used to permit the gain relationships to track in frequency by varying the gain as a function of the local oscillator tuning data. This example shows logic implementation of the skirt-inhibition method as used in early DF systems. Based upon this and developments in radar, an evolution to monopulse passive direction finding took place.

During the late 1940s declassification of World War II documentation describing monopulse radar methods resulted in the publication of Rhodes' "Introduction to Monopulse" (3). Although monopulse had been thoroughly described earlier (4–6), Rhodes' book presented a unified theory of monopulse that became the basis of much of the technology now in present use. Monopulse methods were applicable to the receiver section of radars, permitting the development of the DF angle of arrival and tracking data passively. The associated transmitter emitted nonscanning, constant illuminating signals thus eliminating the mechanical problems of conical and rotating scanners while greatly improving DF accuracy overall.

In the receiver, amplitude monopulse techniques used the ratio of the signals received in two identical but displaced antenna beams as shown in Figure 2-7. The

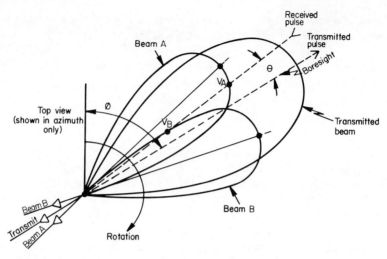

Figure 2-7. Monopulse relationships between a transmit and two receive antennas co-boresighted and rotated together.

radar signal is transmitted, illuminates the target, and is reflected back to the radar antenna. It is received by the two offset overlapped receiver beams, *A* and *B*, which are mechanically aligned to point in the same direction (co-boresighted) and rotate with the transmit antenna. (This is shown for azimuth only in the figure; actual radars develop both azimuth and elevation signals by using a third or elevation offset channel identical to the others.) DF information is fully defined by noting the antenna scan position for coarse angle and the ratio of the voltages in the two receive beams for fine angle, which, as will be shown, is independent of interpulse variation and target glint. The improved accuracy resulting from the ability to remove interpulse variations and the ability to measure DF on each pulse was a significant breakthrough in obtaining high-precision DF data. Since most of the work was done at microwave frequencies at which high gains and dimensionally feasible waveguide devices could be constructed, it was natural to transition monopulse methods to the microwave DF receiver.

The simple amplitude monopulse technique can also be applied to solve the DF receiver lobe-inhibition problem. Figure 2-8 show a monopulse implementation of the previously described dual-channel superheterodyne passive DF receiver. The front-end circuitry is identical, the difference being the method used to determine that the DF signal is greater than the omni. The previously described channel selection logic has now been replaced with analog circuitry that subtracts omni video from DF video *only* when omni video is present and less in magnitude. This is done by inverting the omni video and adding it to the DF signal instantaneously on a pulse-by-pulse basis or by using a two-input operational amplifier. This process may be explained as follows:

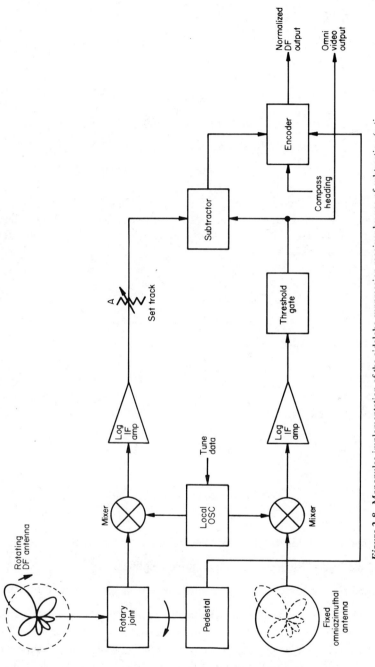

Figure 2-8. Monopulse implementation of the sidelobe suppression receiver by use of subtraction (ratio tracking).

21

Assuming an intercept I,

$$I = T(t) \, \Psi(t) \, f(\omega t) \qquad (2\text{-}1)$$

where

$$\Psi(t) \;\; = \;\; \text{target scan modulation}$$

$$T(t) \;\; = \;\; \text{intrapulse modulation or glint}$$

$$f(\omega t) \;\; = \;\; \text{RF signal}$$

$$\theta \;\;\;\; = \;\; \text{pointing angle of the DF antenna}$$

The signal is received by the scanning DF antenna with a gain of G_p when θ equals the angle of arrival of the signal, yielding $G_p = f(d\theta/dt)$, which contains the desired angle of arrival. The signal is simultaneously received by the omniazimuthal antenna at a constant gain of G_o. The DF signal is

$$G_p(\theta) = G_p \, \frac{d\theta}{dt} \, T(t) \Psi(t) f(\omega t) \qquad (2\text{-}2)$$

and the omniazimuthal signal is

$$G_o I = G_o T(t) \Psi(t) f(\omega t) \qquad (2\text{-}3)$$

Consider now the gain of the logarithmic amplifiers

$$G_1 = m_1 \log r_1 l_1 (1 + A) \qquad (2\text{-}4)$$

$$G_2 = m_2 \log r_2 l_2 (1 + A) \qquad (2\text{-}5)$$

where

$$G_1 \text{ and } G_2 \;\; = \;\; \text{logarithmic amplifier transfer characteristics}$$

$$m_1 \text{ and } m_2 \;\; = \;\; \text{slopes of the logarithmic amplifiers}$$

$$r_1 \text{ and } r_2 \;\; = \;\; \text{base of the logarithmic system, assumed here to be 10}$$

$$l_1 \text{ and } l_2 \;\; = \;\; \text{offset voltages of the logarithmic curve}$$

The argument $(1 + A)$ is present to account for the case of $A = 0$, which would make the logarithm go to infinity (note that the logarithmic curve of Figure 2-5 does not go through the origin.

By definition, the two logarithmic amplifiers are equal and assuming $A >> 1$ we can let

$$G_1 = m_1 \log_{10} l_1 A \tag{2-6}$$

$$G_2 = m_1 \log_{10} l_2 A \tag{2-7}$$

The application of the signal to the logarithmic IF in each channel demodulates (removes) the $f(\omega t)$ carrier resulting in a video output signal.

Letting G_1 be the DF channel and G_2 the omni from (2-1) and (2-2), we get

$$G_1 = m_1 \log_{10} l_1 \left[G_p \left(\frac{d\theta}{dt} \right) T(t) \Psi(t) \right]$$

Letting G_2 be the omni from (2-3) and (2-7) gives

$$G_2 = m_2 \log_{10} l_2 \left[G_o T(t) \Psi(t) \right]$$

the subtraction of logarithms forms their ratio as follows:

$$G_1 - G_2 = m_1 \log l_1 \left[G_p \left(\frac{d\theta}{dt} \right) T(t) \Psi(t) \right]$$

$$- m_2 \log l_2 \left[G_o T(t) \Psi(t) \right] \tag{2-8}$$

$$= m \log \frac{G_p \left(\dfrac{d\theta}{dt} \right) T(t) \; \Psi(t)}{G_o T(t) \; \Psi(t)}$$

reducing (for $G_o = 1$) to

$$\Delta G = m \log G_p \left(\frac{d\theta}{dt} \right) \tag{2-9}$$

assuming $l_1 = l_2$ and $m_1 = m_2 = m$.

Equation (2-9) shows that the output of the subtractor is only proportional to the angle of arrival of the signal, shown here as a function of the rotation of the direction finding beam $d\theta/dt$. *The scan modulation of the emitter has been removed as has the intrapulse or glint occurring during the pulse width of the signal while the skirt response is inhibited, as before.* The DF accuracy is therefore independent of the signal level for a given set of antennas to the extent of an acceptable signal-to-noise ratio, eliminating distortion of the DF beam. It is easy to see why this technique is such a powerful tool.

The skirt-inhibited receiver described here contains one basic element of monopulse technology, namely, normalization, which, as shown, provides cancellation of input signal variation and common terms in the formulation of the logarithmic ratio of the two beams. In Equation (2-8), these terms canceled since they were, by definition, equal ($l_1 = l_2$). The terms other than the input signal variation, due to rotation of the DF antenna, canceled out due to matching, which is only as good as the equality that can be achieved between detectors and logarithmic amplifiers. This is a major factor in the error budget of the system.

2.2 CONCEPT OF MONOPULSE

In the example shown in Figure 2-4, the omnidirectional antenna was used to allow unambiguous isolation of the desired DF beam response with respect to the undesired lobe responses. No attempt was made, however, to improve the accuracy of the DF measurement beyond that of assuring unique recognition. The accuracy of the system is still dependent upon determination of the angle at which maximum signal occurs at the peak or boresight of the DF beam. The use of the omnidirectional antenna in the ratio determination adds no improvement in angular articulation since it has equal response for all angles. In radar and direction finding, high accuracy is the desired essential. Monopulse methods achieve this for radars by replacing the omniazimuthal beam with a second DF beam, as was shown in Figure 2-7, to provide a ratio signal whose amplitude (or phase) permits a vernier determination of where the signal lies within the beam and with respect to the boresight. This higher accuracy derives from the added slope of the second DF beam and from RF methods of angle formation that maximize the relationship of the pattern differences.

Consider a rotating antenna with two beams offset by physically pointing or squinting the feed elements, as shown in Figure 2-7. (The figure shows patterns that can be obtained by pointing the radiating elements directly or by reflecting their patterns from a passive dish or other conducting–reflecting surface.) A signal may be first detected by either beam, which allows the ratio of the beams to be formed at an appropriate signal-to-noise ratio that determines the system false alarm rate. By noting all the angles at which these events occur, an accurate DF measurement is made on each arriving signal or pulse. The angle of arrival is then obtained by calibrating the ratio voltage. In the figure, a return at $\phi - \theta$ clearly sets up V_A and V_B giving angle θ away from ϕ, the boresight position. This improves the accuracy of the simple DF system, described before, because of the greater decibel degree ratio of the DF to omni null formed by the two offset receiver beams A and B. By the use of sum and difference RF techniques, it is possible to develop even deeper nulls for greater accuracy and to obtain a third beam for radiating a transmitting signal (the radar case), all at the same time. This can be done for both azimuth and elevation. By rotating the combinational feed network and its reflecting dish, it is possible to provide both scan and tracking in one antenna assembly, called track-while-scan (TWS).

In the process of understanding the above relationship, we have encountered another monopulse requirement; the need for symmetry about the boresight axis. This symmetry must be odd (skew) to permit determination of which side of boresight the intercept at $\phi - \theta$ lies, since it may be seen that the magnitude of the ratio is the same for θ as for $-\theta$. In the case shown, knowledge that V_A is greater than V_B gives this answer; in more sophisticated systems, the answer must be implicit in the result.

2.2.1 The Postulates of Monopulse

Rhodes (3) gives three postulates for the foundation of a unified monopulse theory. Although they are often considered to reduce to only two, it is useful in our understanding to state all three as follows:

1. *Monopulse angle information always appears in the form of a ratio*. This is the concept of receiving a signal simultaneously in a pair of antennas covering the same field of view and then comparing the signals by forming them into a ratio. The value of this ratio is independent of the signal and any common noise or modulation present in it.

2. *The ratio of a positive angle is the inverse of the ratio of the negative angle*. This postulate requires that the patterns be symmetrical about the boresight axis and when coupled with the requirement for a ratio, implies that both antenna patterns are able to receive the signal, which, for passive direction finding, may not always be the case.

3. *The angle-output function is an odd real function of the angle of arrival*. This last postulate was added for completeness to postulate 2, which considers the general case of the angle of arrival, a complex number. The output of the monopulse system is considered by Rhodes to be a sensing ratio $r(u)$. Postulate 3 indicates that the real output of the detection process is the negative angle measured from the boresight. This is a general-case definition of odd or skew symmetry for complex numbers:

$$\text{real } F[r(u)] = - \text{ real } F[r(-u)]$$

The general theory of monopulse considers that the angle sensing function falls into one of three categories: amplitude, phase, or a combination developed by RF combining, called sum and difference $(\Sigma - \Delta)$.

Amplitude sensing is based on the fact that the angle of arrival is contained in the amplitude ratio of two beams at a common phase center origin that are physically displaced, or squinted, or overlapped to provide a displacement of the patterns about boresight as shown in Figure 2-9. This is also the technique that has been described when discussing Figure 2-7. Phase sensing makes use of two beams that have a physical displacement of their phase centers, usually by several wavelengths. This makes their patterns appear essentially parallel or overlapped in the far field as shown in Figure 2-10. An input signal arriving off boresight will arrive at one beam before the other representing a time or phase angle difference. The difference in the phase angle as measured in each antenna of an arriving phase front of a signal is Ψ, which from Figure 2-10 can be found by considering the difference in path length of $S = D \sin \phi$ due to the antenna aperture displacement D. Letting ϕ be the phase lag due to the difference in the time of arrival of the two signals gives

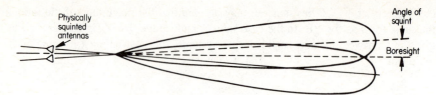

Figure 2-9. Amplitude monopulse response of two physically squinted antennas.

$$\Psi = -2\pi \frac{S}{\lambda} = -2\pi \frac{D \sin \phi}{\lambda} \tag{2-10}$$

where

ϕ = the angle of arrival measured from bore sight

λ = the wavelength

If A and B are the RF voltages at each antenna, then

$$A = M \sin(\omega t) \tag{2-11}$$

and

$$B = M \sin(\omega t + \Psi)$$

$$= M \sin\left(\omega t - \frac{2\pi D}{\lambda} \sin \phi\right) \tag{2-12}$$

where M is a common constant due to transmitted power. This shows that the angle of arrival ϕ is contained in the RF argument or phase difference of the two beams for all signals off the boresight axis. Measurement of this phase difference will

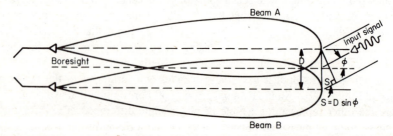

Figure 2-10. Phase monopulse response due to time difference of arrival of signal in two parallel beam antennas.

yield ϕ, the angle of arrival with some ambiguity problems due to frequency of operation, which will be discussed later.

The distinction between amplitude and phase monopulse sensing can be stated as follows: In amplitude monopulse systems, the ratio signal is the result of comparing the amplitudes of two signals from two displaced antenna patterns that essentially originate from a *single phase center* and overlap (have some part of the received signal in common) in the far field; in phase monopulse the ratio is formed by comparing the time difference of arrival of two patterns originating from a linearly *displaced phase center* that overlay (are superimposed) in the far field. In amplitude monopulse systems, the amount of pattern displacement gives angular articulation. In phase monopulse, the difference in time of arrival contains the angle information. In either system, both beams receive the signal.

2.2.2 The Monopulse Comparator

In the unified concept of monopulse, three basic functions are performed: angle sensing by the antennas, ratio conversion by RF sum-and-difference comparator circuits, and angle detection. We have discussed only angle sensing thus far. The description of the amplitude and phase monopulse sensing implies a commonality: one can be readily converted to the other mathematically by adding a 90 degree phase shift (7).

There are three possible outputs from the ratio conversion function. The first is amplitude, where angle information is contained in the detected voltages of the squinted overlapped beams. The second is phase representing the time-difference-of-arrival of a signal received by two parallel overlayed beams. The third is a set of sum-and-difference signals received by two beams that may be used with either of the two angle-sensing methods by combining the signals in an RF comparator or hybrid network. In this case, the angle information is contained in both the amplitude (as a magnitude) and phase (as a sign) of the resulting signals.

In passive direction finding, all three types of systems will be encountered with certain variations not found in texts describing radar monopulse methods. It is necessary to understand the differences between the radar systems and to appreciate the operation of the different passive DF systems, since use is often made of a combination of techniques. Amplitude and phase antenna sensing have been described above; it only remains to discuss the sum-and-difference approach.

The operation of a comparator makes use of the well-known properties of a 3 dB hybrid or "magic T" directional coupler. Figure 2-11 shows the relationship of voltages at the various ports for this device. The coupler shown on the left side of the diagram is a 90 degree quadrature device with the equivalent of 90 degrees of additional phase shift added and subtracted in the path 2–3. A voltage applied at 1 produces an output voltage at 4 in phase and a voltage at 3 in phase shifted $+90$ degrees by the coupler and -90 degrees by the phase shifter ($-j$). A voltage applied at 2 produces an output at 3 shifted 0 degrees in phase due to the cancellation

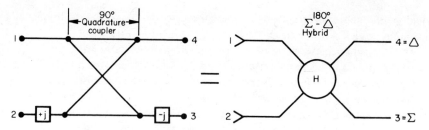

Figure 2-11. Diagram of a 180 degree 3 dB hybrid coupler.

by the opposing $+j$ and $-j$ (90 degrees) shift adding to the $+90$ degree shift in the coupler path from 2–4. As a result, the following relationships prevail:

Port	Voltage	Function
1	V_1	Input 1
2	V_2	Input 2
3	$\dfrac{\sqrt{2}}{2}(V_1 + V_2)$	Sum (Σ)
4	$\dfrac{\sqrt{2}}{2}(V_1 - V_2)$	Difference (Δ)

It may be seen that the sum port (Σ) adds the voltages V_1 and V_2 while the difference port (Δ) subtracts the voltages V_1 and V_2.

The comparator may be connected to either the phase or amplitude angle sensors. Figure 2-12 shows the resultant using the phase antenna connection. The output

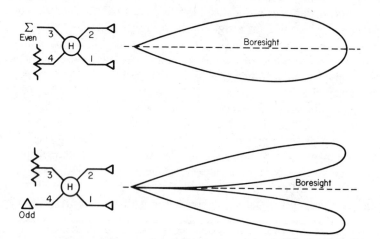

Figure 2-12. Sum-and-difference outputs from a hybrid feed network. (*a*) Sum (Σ) channel pattern. (*b*) Difference (Δ) channel pattern.

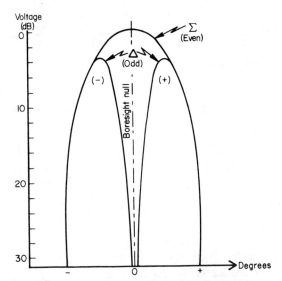

Figure 2-13. Ratio of the sum (Σ) and difference (Δ) patterns.

appearing at the sum (Σ) port is an even symmetry broad-beam pattern centered on boresight. The output of the difference port is an odd symmetry split-beam pattern with a sharp null also centered on boresight representing the odd pattern as shown in the figure.

If the ratio of the sum divided by the difference is taken, the resultant will be a pattern of the form shown in Figure 2-13. It may be seen that an extremely deep unambiguous null capable of high resolution formed at boresight, where there is a phase reversal represented by the change of angle sign on either side. If the output of the amplitude angle sensor is to be connected to the comparator, a 90 degree phase shifter must be added to effectively convert the amplitude voltage difference to a phase angle difference. This latter point is significant since it indicates that *amplitude sensing and phase sensing can be converted from one to the other by a 90 degree shift*, which is often accomplished in practice.

2.3 MONOPULSE ANGLE DETERMINATION

The angle-processing circuitry is designed to form the ratio stated in the first postulate and to extract the magnitude and, if necessary, the sign of the angle. As might be expected, since there are three types of monopulse angle determination methods ($\Sigma-\Delta$), amplitude voltage, and phase angle, there will be three types, or classes, of angle processors for radar systems. For passive direction finding we will add two variants, which for our purposes are perhaps of the greatest interest of all.

The first case, class I shown in Figure 2-14, is the angle processor used with sum (Σ) and difference (Δ) angle sensor voltages as developed from the ($\Sigma-\Delta$)

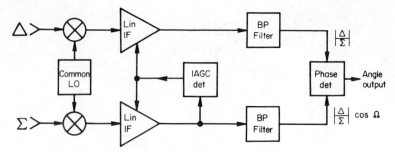

Figure 2-14. Class I sum (Σ) and difference (Δ) monopulse processor.

hybrid. Since the input signal amplitudes and RF phases contain the desired angle-of-arrival information, it is necessary to preserve balance of both phase and amplitude throughout the system. A superheterodyne configuration is shown using linear IF amplifiers in conjunction with bandpass filters that are chosen to establish the optimum bandwidth for the known radar pulse return. The requirement for ratio taking and amplitude balance in this class of processor requires that the IF amplifiers always be operated within a sufficient linear dynamic range, necessitating the use of an instantaneous automatic gain control (IAGC) detector, operating on the sum channel. Common gain control is accomplished by feedback to both channels. This IAGC technique is actually an electronic divider. If one channel of an identical two-channel amplifier is used to develop an automatic gain control voltage that controls the gain of the second channel, then any signal fed to the second channel will be divided by the first (8). This is shown in Figure 2-14, where the angle output of the phase detector is the monopulse ratio voltage magnitude Δ/Σ multiplied by cos Ω, where Ω is the difference in the phase angle between the Δ and Σ channels.

The class I processor, while often used in radar applications, finds little use in broadband electronic warfare (EW) DF receivers since the EW receiver's wider bandwidth, the need for phase balance between the channels, and the limited dynamic range of linear amplifiers impose restrictions on accuracy.

The class II processor shown in Figure 2-15 is designed to operate with amplitude monopulse systems where angle information is fully contained in the detected voltages. A superheterodyne receiver with logarithmic IF amplifiers provides conversion from RF to video, a process that removes all phase information. The extraction of angle data is accomplished by subtraction of the logarithmic voltages, thus providing the normalization ratio. The magnitude of the output is proportional to the angle of arrival, and the determination of which channel is of greater magnitude can be accomplished by the use of bipolar subtraction of the logarithmic outputs. The sign (+ or −) of the output indicates the greater magnitude channel. This type of processor is representative of the types used in current radar systems and finds application in passive scanning superheterodyne DF receivers where ambiguity in defining the greater channel under noisy conditions is offset in part by the high sensitivity and wide dynamic range this configuration provides. Since only amplitude

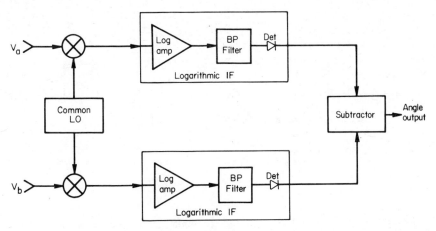

Figure 2-15. Class II amplitude monopulse angle processor.

need be critically balanced (phase unbalance can be tolerated), wide dynamic range IF-type logarithmic amplifiers can be used even though they tend to exhibit some phase variation as a function of signal level.

One variant of the class II processor, the detector logarithmic video amplified (DLVA) type, shown in Figure 2-16, is the most popular type for microwave passive direction finding. This processor works only with amplitude monopulse antenna systems since the initial detection process removes all phase content. The chief reason for DLVA popularity is the wide instantaneous bandwidth obtainable by direct detection of the RF output of a wideband antenna, such as a spiral. The use of this technique will be the subject of several of the following chapters; however, the state of the art of development of matched amplifiers and detectors has advanced to the point where duty cycle (the ability to handle dense pulse signal environments) becomes the major limitation. DLVA units matched to better than ±0.5 dB over 60 dB dynamic ranges are generally available (2). Increased sensitivity of DLVA or "crystal video" type monopulse receivers is now attainable with the availability of broadband RF amplifiers, which have gains of 25 dB, low noise figures, and

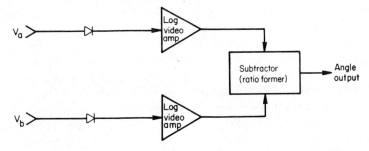

Figure 2-16. The detector log video amplifiers (DLVA) monopulse processor.

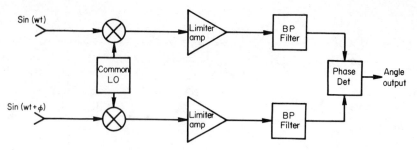

Figure 2-17. Class III phase monopulse angle processor.

bandwidths of nearly 20 GHz. Amplitude matching across the RF band is essential over all conditions in the two channels.

The class III and angle processor, Figure 2-17, is designed for phase monopulse input and requires careful phase matching throughout. It is often used in conjunction with one of the other classes of processors to resolve ambiguities or to give the sign of an angle. Modern hard (sharp knee) limiters have made the class III type of processor popular. It is possible to obtain fairly wide bandpass characteristics with reasonable phase matching. The output angle is a voltage that is a sinusoidal function of the angle of arrival of the form often found in instantaneous frequency encoders. Some systems use combinations of monopulse measurements, which can optimize the design of the computer/processor, allowing application of phase encoding as a by-product. An example of this is the use of a class III processor for phase sign and a class II processor for angle magnitude in a ($\Sigma-\Delta$) system.

As might be expected, there is an analogous wide bandwidth variant to the class III phase processor called the phase correlator, which is a broadband phase-measuring technique analogous to the DLVA. This type of system, also known as a phase discriminator, is shown in Figure 2-18 and is described in detail in Chapter 10, Section 10.2. Two signals $K \cos(\omega t)$ and $K \cos(\omega t - \Psi)$ are hard limited and fed to a printed-circuit four-hybrid network consisting of one 180 degree and three 90 degree hybrid couplers. The DF information, contained in ϕ, appears in the output as either the sine or the cosine of ϕ. The ratio can be taken either by digital or analog (logarithmic amplifier) means to form the tangent and cotangent of the phase difference. The wideband correlator shown here has many applications and is more fully described below. It is a true *phase* monopulse comparator since all amplitude variations are removed as a result of the limiters, which are designed to limit with a sharp knee in the amplitude response curve down to the noise level. The correlator is used chiefly with interferometer phase measurement DF systems and with instantaneous frequency measurement (IFM) receivers, where the difference in phase to be measured represents the difference in path length (phase) of a frequency discriminator. The output sine and cosine values in this case represent the frequency of the incoming signal. Again, all amplitude variation is removed before the frequency discriminator to make the output insensitive to input variation (normalization).

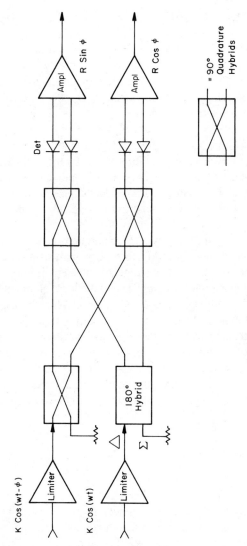

Figure 2-18. The wideband class III phase correlator.

33

2.4   BIRTH OF PASSIVE DIRECTION FINDING TECHNIQUES

The birth of microwave passive direction finding has been shown to be an evolution from its parent, radar. In early radar, a modulated signal was transmitted, reflected back from the target, and received by the same antenna, giving range and angle-of-arrival data as determined from the angular position of the radar antenna. The need to improve this positional information gave rise to the concept of tracking (or track-while-scan), a method making use of the *difference* in the transmitted and received modulation to provide a feedback signal to track the target. Monopulse methods provide high-precision angular information by the positional or phase offset of the received beams allowing the radar to transmit an unmodulated constant amplitude signal. At this juncture, the concept of passive direction finding was born: Eliminate the radar transmitter and use the signals transmitted by the target for its own purposes to provide the "return."

From a system viewpoint, the type of configuration to be used for passive direction finding depends upon many factors. Any analysis must consider the purpose of the system—warning or intelligence; the environment to be encountered—strategic or tactical; the host vehicle and the location of antennas—whether the antennas are to be single point mounted or geographically dispersed. Last, but not least, cost must be considered. If a capital ship is to be protected, a complex antenna array system for ELINT and warning may be justified to best deploy jammer and other ship-protection assets. If small aircraft or helicopters are to be warned, their limited capability to take defensive measures could require a less costly system. A high-performance expensive aircraft would logically require expensive, flush-mounted antenna DF and jamming systems with extensive response capability.

2.5   SUMMARY

The radar evolution to passive direction finding has been presented as an introduction to the solution of generic problems to be encountered. No attempt has been made to present the vast comprehensive bank of literature outlining radar monopulse technology, although the bibliography will lead the reader to detailed analysis of radar monopulse techniques for noise, phase, amplitude, and target error minimization. We shall give the transitions from and develop these relationships for passive direction finding, which makes use of these radar techniques using both monopulse and non-monopulse methods.

The development of passive DF techniques depends in great part upon many considerations, which in turn dictate the systems configuration and parameters. In the following chapters, we shall describe the antenna or primary sensor, showing how in many ways this dictates the system design. Descriptions of the simple antenna types will evolve into systems for their use, showing how certain types and classes of operational systems have become standard.

REFERENCES

1. Howells, P. W., "Intermediate Frequency Sidelobe Canceller," U.S. Patent 3,202,990, Aug. 1965.
2. Lipsky, S. E., "Log Amps Improve Wideband Direction Finding," *Microwaves*, May 1973.
3. Rhodes, D., *Introduction to Monopulse*, New York: McGraw-Hill, 1959.
4. Budenbom, H. T., "Monopulse Tracking and the Thermal Bound," *IRE National Convention on Military Electronics*, June 1957.
5. Page, R. M., "Monopulse Radar," *Convention Record Institute of Radio Engineers*, No. 55.
6. Page, R. M., "Accurate Angle Tracking by Radar," NRL Report RA 3A222A, Barton, D. K., *Monopulse Radars*, Dedham, MA: Artech House, 1974, Vol. I.
7. Rhodes, *Introduction to Monopulse*, p. 29.
8. Kirkpatrick, G. M., "Final Engineering Report on Angular Accuracy Improvement," General Electric Report, 1 August, 1952, Barton, David K., *Monopulse Radars*, Dedham, MA: Artech House, 1974, Vol. I, p. 77.

Chapter Three ───────────────

Antenna Elements for Microwave Passive Direction Finding

The primary source of DF accuracy is the antenna. We will therefore study the various types of antennas used for microwave passive direction finding. There is almost an infinite variety of designs; however, the types to be covered here will be those used in currently deployed systems, as opposed to obscure designs and older types. The antenna, by our definition, is the elemental radiator (or receptor) that can be dispersed, arrayed, or used alone or in combination to achieve specific and predictable angular gain patterns that permit direction finding information to be obtained. A simple example of this definition is the use of a half-omniazimuthal (180 degree) antenna as a side-lobe inhibitor for an array of horn antennas, the gain pattern in this case being tailored to perform this specific function. More complex examples will follow.

Where applicable, each antenna will be studied in its principal mode of operation followed by a discussion of appropriate alternate modes. It is often possible to make a single antenna simultaneously operate in multiple ways to form useful beams for multiple-channel systems in order to improve accuracy or achieve special coverage, the attractiveness being the contiguous field of view (all modes are looking at the same space with the same antenna properties).

3.1 TERMINOLOGY AND CONVENTIONS

The standard IEEE terminology for antennas has been followed throughout this book with few exceptions. Detailed definitions and test procedures may be found in Ref. 1. Figure 3-1 is a diagram of the specific antenna conventions to be used. All of space is described by a Cartesian coordinate system of three planes $X-Y$, $Y-Z$, and $X-Z$, the $X-Y$ plane being assumed horizontal to the Earth. The antenna is situated at the origin of the coordinates and, for the purpose of this discussion, may be considered to be a simple dipole parallel to and coincident with the Z axis and situated in the $X-Z$ and $Y-Z$ planes, causing its far-field radiation pattern to

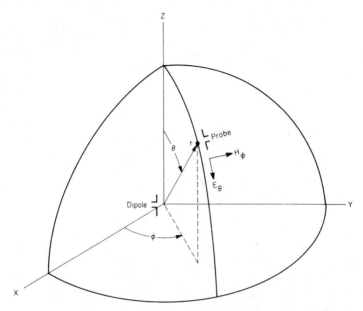

Figure 3-1. Antenna convention.

be directed from the angle θ measured from *Z*. A probe dipole at a fixed distance will, if made to trace all space from the origin, describe a sphere of radius *r*, one point of which is shown in the figure. The fields at this point consist of an electric field E_θ and a perpendicular magnetic field H_ϕ. Since the dipole lies in the *Z* direction, any measurement at a fixed point rotated about the *Z* axis will not change with φ but will change with θ since the *E* field is directive with θ, varying as the sin θ since there is no radiation along the axis of the simple dipole. If we consider the pattern about *Z*, it is a doughnut-shaped torus with an infinitely small hole. The pattern is omniazimuthal (360 degrees) in the φ direction and circular in elevation (consider a cross section cut of our doughnut).

Any plane such as *Z–Y* or *X–Z* that contains the *Z* axis will give the same value of E_θ as the probe dipole is rotated in a circle θ about *Z* at the fixed distance *r*. Since the *E* field vector lies in these planes, they are called *E*-plane patterns. Since the magnetic field *H* is perpendicular to the electric field, a pattern taken in any plane such as *X–Y* is called an *H*-plane pattern. In the convention commonly understood, the *X–Y* plane is assumed parallel to the horizon, therefore *E* field measurements centered about *Z* as a function of φ are referred to as angle-of-arrival directions. Patterns perpendicular to the horizon are *H*-plane patterns taken in the θ direction and are referred to as elevation patterns.

The polarization of a signal is determined by the direction of radiation of the *E* field. If this field lies fully in a fixed plane and is parallel to the *Z* axis, it is called vertical. It is referred to as horizontal if it is parallel to either the Y or X axis. A signal is said to be linearly polarized if the E field varies along a line, which happens if we were to move up and down along the length of the dipole, the field being a maximum at the feed point and zero at each dipole tip. If the E field is constant but varies periodically with time, it is said to be circularly polarized. Right-hand circular polarization sense is defined as the curl of the fingers of the right hand, representing the instantaneous E field as the thumb is pointed in the direction of propagation. If the E field vector does not change in amplitude as it rotates in time, it is perfectly circularly polarized. If it reaches a maximum at some angle of rotation, it is elliptically polarized. The ratio of the maximum to minimum is called the axial ratio and is usually expressed in decibels.

When measuring an antenna field, the polarization of the antenna under test must be the same as that of the radiating antenna, this condition being defined as matched polarization. Any difference in polarization will cause a cross polarization loss, which may be substantial (theoretically infinite) as in the case of right-handed (RH) circular pattern when compared to left-handed (LH) circular pattern or when a vertical pattern is compared to horizontal linear pattern. If an antenna is slant polarized, its E field vector lies 45 degrees from the Z axis resulting in a 3 dB (half-power) loss when measured by a vertical or horizontal antenna, as will the measurement of the field of a linearly polarized antenna when measured by a circularly polarized antenna.

Circular and slant polarization DF antennas are commonly used for microwave DF systems since the intercept's polarization is usually known. There is, of course, the possibility of a complete polarization loss if the sense of the two circular polarizations are different or the slants of the linear polarization differences are 90 degrees opposite as described above. The polarization of the principal lobe of an intercept may differ from that of the side or backlobes, giving the possibility of further gain or losses in the detection capability of the DF. Gain of an antenna refers to the power contained in the 3 dB or half-power beamwidth and is the value of the maximum radiation intensity in a given direction referenced to an isotropic (theoretically lossless) antenna that radiates spherically in all directions. Gains are usually expressed in decibels of gain above an isotropic radiator (dBi) assuming matched polarization. (Gain is sometimes referenced to an ideal dipole although this is not usual.)

When multiantenna systems, such as dispersed spirals or horns, are used to make angle-of-arrival determinations, it is assumed that the polarizations of all of the measuring antennas are equal. If this is not the case, the angle determination may be in serious error. The degree of equality is usually specified by the axial ratio differences between antennas in a given set; hence control of axial ratio is important. Since for many antenna types, axial ratio can vary as a function of the angular distance from the peak of the beam, serious DF errors can result when antenna monopulse comparisons are made off the principal beam axis. These errors become a function of the angle-of-arrival with respect to the DF systems pointing direction and must be treated in the channel-to-channel unbalance error budget. It is often desirable to be able to measure the polarization of an input signal in this case or when high DF accuracies are required.

3.2 SPIRAL ANTENNA

The spiral antenna, which first appears in the literature in the 1956–1961 period (2), is one of the most useful and popular antennas for microwave direction finding since it exhibits wide RF bandwidth and a relatively constant bandwidth beamwidth, is circularly polarized, and can be easily spatially deployed. It is important to understand these general features and the types and modes of the spiral antenna for a full understanding of the usefulness of this type of antenna.

The spiral antenna obtains wide frequency coverage by following the principles of slow-wave or periodic type structures. Simply stated the concept relates to designing an antenna that contains, within its design structure, a radiating dipole pair that is essentially frequency self-selecting by the application of an exciting signal within the frequency capabilities of the antenna. The antenna can perhaps best be understood by imagining sets of dipole antennas of specific lengths connected in series, starting with the highest resonant frequency antenna and ending with the lowest. They are fed by a two-wire transmission line at the high-frequency end, at feed points A and B as shown (for a planar spiral) in Figure 3-2. Any signal with a 180 degree out-of-phase current applied by the transmission line to the antenna will seek out the dipole pair antenna that resonates, or brings the currents in phase, at the signal frequency passing from the highest frequency antenna pair $d_1 - d_1'$ toward the lowest frequency antenna pair $d_N - d_N'$. By resonance we mean the point where the original out-of-phase current flows in phase as a result of the phase change caused by propagation over the *differential* distance between the dipole pair and the feed point. This differential distance assumes that one arm of the pair excites

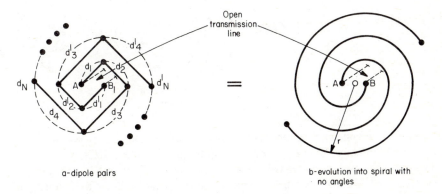

a-dipole pairs   b-evolution into spiral with
                                                        no angles

Figure 3-2. Evolution of series connected dipoles into the archimedian spiral antenna. (*a*) Dipole pairs. (*b*) Evolution into spirals with no angles.

an adjacent arm, which when compared to the first arm can be evolved into a circular spiral by bending the finite dipoles, eliminating the d chords, and replacing them with an arithmetic spiral with an increasing radius r as shown.

The spiral antenna can be constructed as a pair of metallic radiators or as a pair of slots that are self-complementary, that is, the width of the spirals are equal to the spacing between them. This concept derives from Babinet's optical principle that stated that if light were diffracted by a screen resulting in light patterns that are the complement of the screen pattern, then the sum of the screen and its complement would be the pattern without the screen. This is analagous to placing a photographic negative over a contact print aligned so that no pictures can be discerned. In field theory (3), a radiating conductor is the complement of a conducting sheet that has a slot removed that is exactly equal to the shape of the conductor. If the magnetic field present in the conductor is compared with the electric field in the slot and if line integrals are performed for each to determine the impedance, the result for a two-arm structure is

$$Z_C Z_S = \frac{n_o^2}{4}$$

where

$$Z_C = \text{input impedance of the radiating conductor}$$
$$Z_S = \text{input impedance of the slot}$$
$$n_o = 120\pi, \text{ the impedance of free space}$$

then for the self-complementary case

$$Z_C = Z_S = Z_{\text{spiral}}$$

$$Z_{\text{spiral}} = \sqrt{\frac{n_o^2}{4}}$$

$$= \frac{n_o}{2}$$

$$Z_{\text{spiral}} \cong 189 \text{ ohms}$$

A spiral antenna can be constructed to cover a frequency bandwidth of over 36 to 1 since the above impedance equation is frequency independent, resulting in a design that has well-behaved voltage standing wave ratio (VSWR) and beam characteristics. The geometry of the construction and the relationship of the spiral dipoles gives rise to a 90 degree phase shift of the E vector with time, yielding a circular polarization response characteristic. This is a highly desirable attribute for a DF antenna since circular polarization is responsive to the four forms of possible emitter polarization: horizontal, vertical, RH circular, and LH circular. A single spiral will only respond to the circular polarization response matched to it; however, both RH and LH responses can be received by certain antenna designs. Although dipole and other types of resonant antennas have many of the features of spirals, few offer the wide bandwidth and circular polarization characteristics attainable with this antenna. The frequency of operation of the spiral is usually defined as the bandwidth over which the spiral exhibits circular polarization of the principal beam.

The beamwidth of a typical spiral, when excited in the principal mode, is fairly constant at approximately 70 degrees, typically increasing at the lowest frequency of operation. This is a good number for geographically dispersed antennas, since only four or six antennas are required to cover 360 degrees of azimuth. A specific spiral design, the cavity-backed planar type, can be manufactured by photoetching copper-clad dielectric materials using printed circuit techniques, yielding a high degree of uniformity. This allows good phase match between units and is ideal for interferometry and phase monopulse applications.

There are basically three commonly used types of spiral antennas: the planar archimedian or arithmetic spiral, the planar equiangular spiral, and the three-dimensional conical equiangular spiral, shown in Figures 3-3, 3-4, and 3-5, respectively. Each type has several variants: The circular spiral antenna can be constructed in a square configuration, which for some interferometer systems permits closer spacing, the planar equiangular spirals can be constructed with differing offset angles for different tracking results, and the conical spiral can be designed with narrow apex angles to permit stacking two as a pair such that the electrical phase center distance between antennas can be made less at the high-frequency end of the range, a desirable feature for interferometer DF systems.

Figure 3-3. Cavity-backed circular planar archimedian spiral antenna. Courtesy of AEL.

Figure 3-4. Circular planar equiangular spiral antenna. Courtesy of Transco.

3.2.1 The Planar Archimedian Spiral

The planar archimedian spiral antenna, shown in Figure 3-3, consists of a photoetched copper Duroid or type G10 dielectric material printed circuit that radiates bidirectionally, perpendicular to the plane of the copper winding. For narrow-band high-gain applications, the antenna circuit board can be backed with a metallic cavity. For wideband applications, anechoic absorbing material is used in a backing cavity to

Figure 3-5. Conical, three-dimensional spiral antenna. Courtesy of AEL.

absorb the back wave completely, which is analogous to audio frequencies where the back wave of a loudspeaker is absorbed by an insulated cabinet, at the expense of 3 dB of gain. The figure shows the construction of a commercially available two-arm broadband circular spiral covering the 2–18 GHz range.

The two-arm spiral antenna has been considered (4) to be a balanced two-wire transmission line that gradually is formed into an antenna by the action of the increasing radius of an archimedian spiral:

$$r = r_o e^{a\phi} \qquad\qquad (3\text{-}1)$$

where

$$r_o = \text{the radius at the start}$$

$$a = \text{the growth rate}$$

$$\phi = \text{the angle position of } r$$

Figure 3-6 shows the relationships of circular and rectangular two-arm spirals and will be used to expand the previous explanations of its operation. The two-wire balanced transmission line does not radiate at the feed point since the currents are out of phase by 180 degrees causing cancellation. As the transmission line A–B forms into the two-arm spiral at some point C, r distance away from the origin 0, there will be another point D, having in common the fact that C and D both lie on a circle of the same radius r centered at 0 and on the straight line $D0CC'$. Point C' can also be considered to be at essentially the same radial distance from the origin as point D if the turns of the spiral are closely (equidistantly) spaced. It is assumed that point C' is excited by C due to its adjacency. The arc distances AD and BC are equal; however, the arc distance BC' is greater. The points essentially

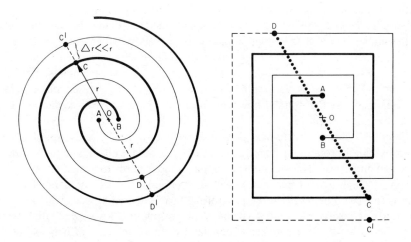

Figure 3-6. Archimedian circular and rectangular spiral relationships.

lie on a circle of radius r measured from the origin 0, therefore, the arc difference BC' from BC is the same as one half of the circumference ($C = \pi 2r$) or πr. When the radius r is equal to $\lambda/2\pi$, the circumference is λ, point C' undergoes an additional phase change of $\lambda/2$ or 180 degrees with respect to point B due to the differential length ($\pi r = \lambda/2$ when $r = \lambda/2\pi$) and the 180 degree antiphase current condition originating at the feed point is now an in-phase condition at $B-C'$. This causes radiation to take place, which only occurs for in-phase conditions. A similar in-phase condition occurs at $D-D'$. The radiation occurs in n annular rings or bands of circumference $n\lambda$. Since the spiral radius is continuously growing and since the spacing between the conductors equals the width of the conductors, the antenna is a broadband device. Application of a signal at a frequency within the limitation that its wavelength is less than the maximum diameter will essentially select the appropriate radiating band. Antennas of this type are referred to as slow-wave since the signal propagates to where it can radiate.

The spiral antenna is circularly polarized; for every differential group of elements that have shifted 180 degrees in phase at the diameter of radiation, there is another group that is in time and space quadrature since the phase of the groups is varying as a function of the spiral growth rate. This causes a 90 degree phase shift making the spiral response circular. The circularity of the response is excellent at bands within the outer or largest ring degrading to linear polarization near the low-frequency or outer band edge and at the high-frequency end if radiation takes place at the connection of the feed line to the antenna structure.

The number of turns in the spiral is not important as long as the spacing between them is small or equidistant for the complementary case. The lowest frequencies radiate from the outer or maximum diameter of the spiral; the highest frequencies, near the center. For most applications the outer edges of the spiral should be resistively terminated to ground to prevent a reverse or reflected signal.

In the square spiral shown in Figure 3-6, the same conditions described above prevail: Point C' is $4l$ greater in length from point C, where l is defined as the perpendicular distance from the origin to a side of the square and where it is assumed the spacing between the turns Δl is constant and much less than l. The circumference of the circle drawn at 0 is approximately $8l$. As above when l is $\lambda/8$ the phase change in current is $\lambda/2$ or 180 degrees, and radiation takes place (each side of the square is $\lambda/4$). There is a 90 degree phase change for adjacent points, providing circular polarization. Square spirals can, as discussed above, be placed closer together than circular spirals; however, phase tracking between square spirals is more difficult to achieve.

Figure 3-7 shows a rectilinear pattern of a cavity-backed circular spiral antenna taken by axially rotating a linearly polarized feed horn while azimuthally rotating the spiral under test. The variations or ripple in the patterns represent the axial ratio response, which may or may not peak in the vertical or horizontal plane. The axial ratio is low (well behaved) at the peak of the beam, degrading somewhat at angles about 60 degrees on either side. The back-lobe response is low (>-20 dB), which shows why this type of antenna is popular for direction finding: It is essentially

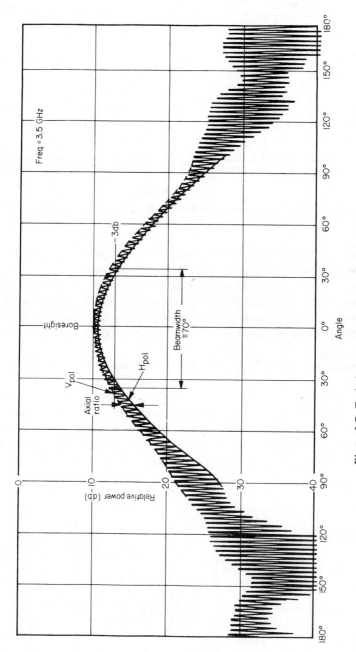

Figure 3-7. Typical pattern of a cavity-backed spiral antenna.

45

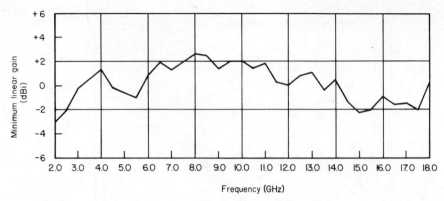

Figure 3-8. Typical antenna gain versus frequency for a cavity-backed spiral antenna.

unambiguous. It is important also to note that there is practically no squint; the peak of the beam occurs in the direction of physical boresight.

Figures 3-8 and 3-9 show the gain and beamwidth of the antenna as a function of frequency over the 2–18 GHz frequency range. The antenna shown here is typical of the type of response obtained in volume manufacturing, which is sufficiently reproducible to allow the spiral to be used for dispersed parallel comparison DF systems.

The operation of the spiral antenna as presented here is often referred to as the "band" theory since it assumes that for every band of circumference of $n\lambda$, where n is a positive odd integer, there will be an in-phase current relationship permitting radiation. The radiation of a spiral fed by a current antiphase transmission line is called the odd mode radiation. When n is an even number, the currents in the evoluting transmission line are antiphase, and a null will result. This theory is conceptually acceptable although not completely subject to mathematical proof.

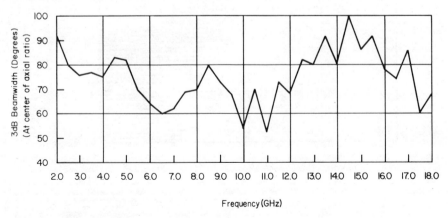

Figure 3-9. Typical azimuthal beamwidth for a cavity-backed spiral antenna.

Each mode of radiation exhibits different characteristics. In the first odd mode $(n = 1, c = \lambda)$, the radiation is perpendicular to the plane of the spiral and at a peak along the principal axis. This mode, which occurs for the diameter, is λ/π $(C = \lambda = \pi D = \lambda/\pi)$. If the spiral is large in diameter and currents exist past the first ring, additional radiation can take place at the third or fifth and higher modes.

The even modes of radiation of a spiral can be excited by feeding the spiral elements in phase. Assuming the same reasoning as above, radiation will take place at even diameters, the principal mode being $C = 2\lambda$. Since the currents started out in phase, in this mode they will be out of phase at $C = \lambda$ and the radiation pattern will exhibit a null. As the line currents progress, they will be in phase at the even mode and radiation will occur. This type of spiral antenna is called a normal mode design and is characterized by a radiation pattern that always exhibits a null along the axis and two peaks, displaced in angle as a function of the frequency, as shown in Figure 3-10.

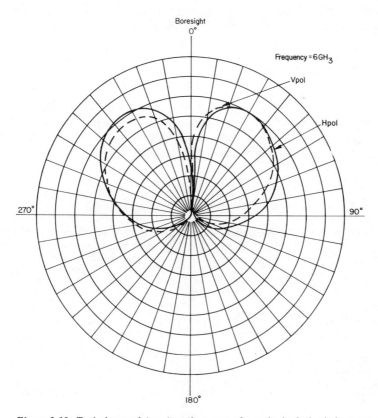

Figure 3-10. Typical normal (even) mode pattern of a cavity-backed spiral antenna.

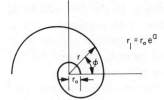

Figure 3-11. Curve for an equiangular spiral.

3.2.2 The Planar Equiangular Spiral

Another solution to the need for wide RF bandwidth antennas is the planar equiangular spiral, first suggested by Rumsey (5) and later characterized by Dyson (6) in 1959. The concept was prompted by the concept of a frequency independent antenna. Work on the logarithmic periodic antenna by DuHammel and Isbell (7) added impetus to the frequency independence theory, as did the development of the archimedian spiral described above.

The essential idea was to build an antenna using a radiating structure that changed dimensions in a linear proportion to the wavelength. If the shape of the radiating element could be defined entirely by angles, the frequency coverage could in theory be infinite for a lossless structure, since angular variation is unbounded. To define a practical finite size, however, one length must be specified. Figure 3-11 shows the development of the idea. If a mathematical spiral is defined as shown, then

$$r_1 = r_o e^{a\phi} \tag{3-2}$$

Any point r from the origin is proportional to the initial starting point at r_o multiplied by an exponential raised to $a\phi$, where a is the growth rate constant and ϕ is the angular position (as specified in the antenna definitions). Letting r_1 define one edge of a conductor, the second edge r_2 can be defined by the equation

$$r_2 = r_o e^{(a-\delta)} \tag{3-3}$$

where r_2 is the outer edge of the conductor of inner edge r as shown in Figure 3-12. The conductor has been constructed by letting $\delta = 90$ degrees. A second

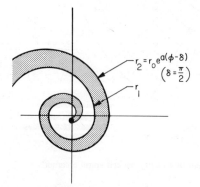

Figure 3-12. Curve for a spiral conductor.

conductor can be added by a similar construction; by letting $\delta = \pi/2$ to make the structure self-complementary, we have

$$r_3 = r_o e^{a(\phi - \pi)} \tag{3-4}$$

$$r_4 = r_o e^{a(\phi - \pi - \delta)} \tag{3-5}$$

The lowest frequency of operation is determined by the maximum radius R resulting in Figure 3-13, showing the complete pattern of an equiangular planar spiral. The radius of the equiangular spiral increases as any conductor goes through 360 degrees or

$$r = \frac{r_o e^{a(\phi + 2\pi)}}{r_o e^{a\phi}} = e^{a 2\pi} \tag{3-6}$$

In general, the Δr is about four with only one and one-half to two turns required to wind a broadband (10:1) spiral. The antenna in the figure is self-complementary and, in fact, was constructed as a slot in copper by Dyson (6) in his early work.

Improved performance of the spiral can be obtained by compound winding; that is, the feed point starts out as an archimedian spiral, transitions into an equiangular spiral, and finally transitions again into another archimedian spiral, which is ultimately resistively terminated at the outer diameter. This is shown for the outer transition

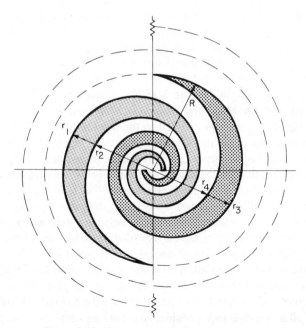

Figure 3-13. Two-arm equiangular spiral antenna.

by the dashed lines in Figure 3-13. This technique improves the bandwidths of the equiangular spiral by reducing radiation effects from the feedpoint and reflections at R due to discontinuities in the termination. Ohmic losses at the low frequency are also less since there is less resistance present before reaching the archimedian spiral. In general, the equiangular spiral exhibits independence to the offset angle δ and the growth rate a; if not, many turns (approximately 3) are used. These few turns present higher VSWR ratios over the typical 9:1 bandwidths (2–18 GHz) than spirals used in most DF systems. The use of the compound winding technique is a method of improving performance, especially the low-frequency band edge. Patterns and gains of the equiangular spiral are similar to those of the archimedian spiral.

3.2.3   *Planar Spiral—Feed and Construction*

The planar spiral antenna is unique in its attainment of performance qualities, desirable for dispersed antenna and phase measurement type systems. Reproducibility must be excellent to permit repeatable known ratios to be established for DF determination and antenna field replacement. The methods of feed and design of the cavity-backing absorber are two additional contributing factors.

RF signals in the popular 2–18 GHz frequency region are usually fed from the antenna to the receiver by 52 ohm coaxial cable, requiring the spiral antenna feed network to effect an impedance transformation of approximately 4:1 (189/52 = 3.6:1) and to develop an unbalanced to balanced transformation required for the 180 degrees out-of-phase feed. This can be done by one of two methods: the balun transformer or the tapered coaxial feedline. In most applications, requirements for axial ratio and squint performance dictate the first method, which has superior performance although it is somewhat more expensive. Figure 3-14 is a cross-sectional view of a typical balun/transformer-fed spiral antenna showing the essential elements starting with the unbalanced coaxial feed, the balun, the tapered line impedance transformer, and the radiating structure all mounted in a metallic enclosure filled with absorbing material. Figure 3-15 is a detailed view of the balun, which can be recognized as the Marchand design (8), with modifications (9).

The input signal is coaxially coupled to the tapered line transformer, which is a small dual-sided printed circuit board with a parallel tapered line printed on each side. The outer shield of the input coax is soldered to one printed tapered line; the inner coax lead passes through the tapered line and is solder connected to a conducting stub, which is dielectrically insulated and mounted in a plug passed through the far wall. The outer wall of the plug is soldered to the outer tapered line. The plug and the stub constitute a wideband open circuit matching stub, which, when viewed at the connection point, permits the current on the shielded input coaxial cable to divide equally between each side of the balanced tapered line. This is due to the fact that the stub and plug constitute a short circuit at the center conductor connection due to the open circuit $\lambda/4$ termination of the stub in the plug, assuring that all of the coaxial center line current flows into the side of the tapered line soldered to the stub. Almost all of the coaxial outer shield current flows into the other tapered line;

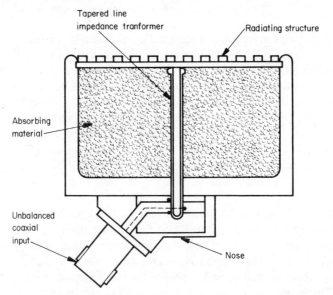

Figure 3-14. Spiral antenna—cross-sectional view.

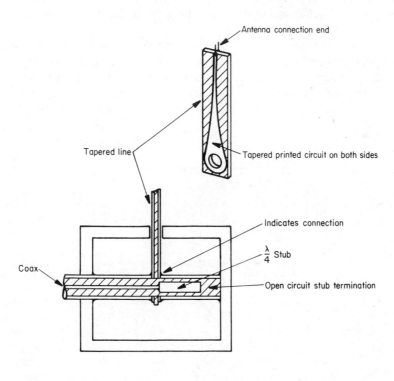

= Dielectric

Figure 3-15. Marchand balun used to feed a spiral antenna (cross-sectional view).

thus an unbalanced conversion is achieved. Good balance is obtained by adjustment of the dimension of the stub and by minimization of the thickness of the tapered line printed circuit. Wide bandwidth is the result of the very good stub short circuit due to the ability to obtain a highly reflective termination at the open end. Although balance is obtained, the necessary impedance transformation is not. The 4:1 tapered line transformer performs this function by converting the 52 ohm input line to the 189 ohm impedance of the spiral. The antenna radiating structure, which is a circuit printed on G10 or Duroid dielectric laminate, is carefully soldered to the tapered end lines to maintain a minimum mass at the connection that might radiate by itself. The entire unit is assembled and backed up by RF anechoic material to fully absorb the radiated back wave.

The split-taper feed, which offers the advantage of low cost, utilizes a direct connection from the coaxial input line to the tapered line without the balancing transformer, as shown in Figure 3-16. A coaxial line has been partially etched away along length L, effecting both an impedance transformation and a balun action. This type of feed depends upon splitting of the current at the outer conductor and division at the septum point as shown. The overall symmetry of the cavity develops a current distribution that achieves a reasonable balanced line effect at the feed point. Although simple, spiral feeds of this type usually display considerably greater pointing error (squint) and axial ratios compared to the true balun-transformer approach.

A normal mode spiral feed is shown in Figure 3-17. Since it is desired to feed both radiating elements of the spiral in phase, they are connected together and fed by the center conductor of the coaxial line. A launcher, positioned behind the spiral radiating structure, is connected to the outer coaxial conductor to achieve balance.

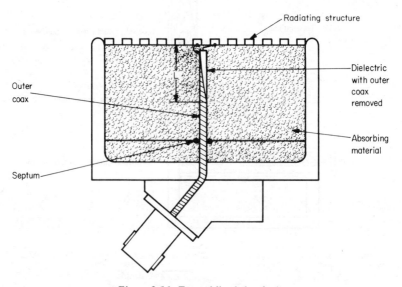

Figure 3-16. Tapered line balun feed.

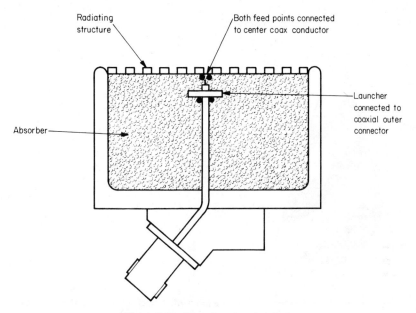

Figure 3-17. Normal mode spiral feed.

It is more difficult to obtain low VSWR over broad bandwidths with this type of feed; however, antennas have been built with predictable performance over 9:1 frequency ranges. Normal mode operation can also be achieved by use of a multiarm spiral and an associated feed network, which is generally a more popular method since normal mode operation of a spiral is usually associated with comparison with other derived spiral beams, all of which are obtainable in multimode configurations.

3.2.4 Conical Equiangular Logarithmic Spirals

The conical logarithmic equiangular spiral antenna results directly from the projection of a planar equiangular spiral on a cone. From Figure 3-18, this projection may be expressed as

$$l_1 = e^{(a \sin \theta_o)\phi} \tag{3-7}$$

where

l_1 = one edge

θ = one-half the apex angle

a = the wrap rate

δ = the spiral width (usually $\pi/2$ for the self-complementary case)

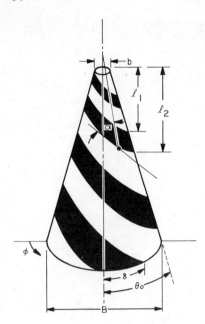

Figure 3-18. Conical equiangular spiral antenna.

The second edge, l_2, may be defined as

$$l_2 = e^{(a \sin \theta_o) (\phi - \delta)} \tag{3-8}$$

The second arm of the spiral is developed by rotation and construction 180 degrees away. All edges of the spiral hold a constant angle α with a radial line, that is, the radiator elements are tightly wrapped around the cone.

As in the case of the planar spiral, the antenna must be fed in current antiphase to excite the odd mode of radiation. Even mode radiation can be excited by parallel in-phase current feed, resulting in a null along the axis. The impedance of a self-complementary spiral is approximately 165 ohms (10). Radiation from the antenna in the commonly used odd mode is off the axis in the pointing direction of the cone ($-Z$ direction) and is circularly polarized degrading to elliptical polarizer at the beamwidth extremes. The lowest frequency of operation is determined by the base diameter B when it equals three-eighths wavelength. The highest frequency of operation occurs when the truncated cone diameter b is one-fourth wavelength. Typical manufactured conical spirals (11) exhibit 0–2 dB gains and cover up to an 11:1 bandwidth using apex angles of about 30 degrees. Higher gains can be obtained by mounting the conical spiral on or below a recessed ground plane, the latter technique often being used for flush-mounted applications. When the apex angle (equals twice θ_o) becomes $\pi/2$, the conical spiral evolves into the planar spiral, which is the limiting case. Unlike its counterpart, however, the conical spiral tends to radiate unidirectionally against a ground plane, eliminating the need for anechoic absorbing materials.

   Since the process of radiation follows the previously described "band" theory,
the conical spiral radiates at what is known as the "active region" with a center of
phase that positions itself linearly along the antenna axis as a function of frequency,
permitting two conical spirals to be pointed toward each other in an array, as shown
in Figure 3-19. This array is very useful for interferometer applications since a
wider unambiguous bandwidth can be obtained as a result of the closer spacing
($d < \lambda/2$) at the high-frequency band edge. By interconnection with a 180 degree
hybrid, sum and difference monopulse outputs can be obtained over an extremely
large bandwidth. An extensive study of this approach by Hahn and Honda (12)
reached the conclusion that practical DF systems could be made, although with
only a fair degree of accuracy, due in part to the phase tracking of componentry
available at that time. The more popular configuration for conical spirals is in
amplitude comparison monopulse systems where the greater gain of the conical
spiral is important.

3.2.5   Multimode and Multiarm Operation of Spirals

A single spiral can be made to appear as an array of many to provide multiple
outputs for different types of monopulse DF systems. This can be done by exciting

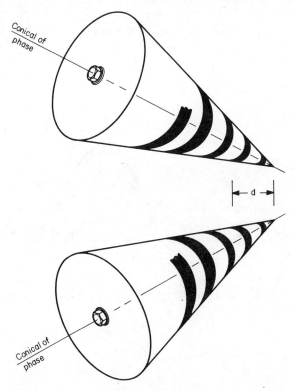

Figure 3-19. Center of phase of two pointed conical spirals d, moves with frequency $d = f(wt)$.

multiple modes of operation in a simple two-wire spiral or by the use of multiarm spirals. Both cases require the use of combinations of couplers and hybrids, which, because of their narrow bandwidths, limit the overall bandwidth obtainable. This type of use is important, however, since it is possible to develop highly accurate targeting and guidance systems in a small space utilizing a minimum of antenna "real estate," often a key consideration in missile and spacecraft applications.

First attempts to excite dual modes in a simple two-arm spiral were described by Kaiser et al. (13) in a passive type direction finder. The concept was to excite both the even and odd modes simultaneously in a two-arm spiral by means of a hybrid, recognizing that there would be two distinct patterns generated. The first, due to odd mode excitation, would be the familiar beam pointed along the axis perpendicular to the plane of the antenna. The second would be the even or normal mode pattern, a doughnut-shaped torus with a null along the axis. Figure 3-20 shows the phase relationship between the two patterns as a function of revolution around the Z axis or circumference of the spiral. The odd or first mode angle ϕ travels 360 degrees, while the even or second mode angle traverses 2ϕ or 720 degrees. The measurement of the *difference* in phase between the two modes therefore will give the angle ϕ. The amplitude difference between the beam and the torus provides θ. Knowledge of these two angles provides an unambiguous measure of the angle of arrival of a signal.

Although conceptually attractive from many points of view, this system suffered limitations due to phase variations at the extremes of the beams and bandwidth

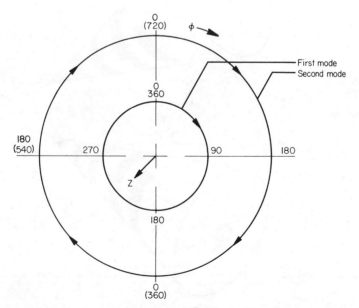

Figure 3-20. Phase relationships of a single-arm dual mode excited spiral antenna.

limitations due to the lack of wideband hybrids at that time. On the positive side, there is the potential for better ellipticity tracking since both modes use the same antenna. The possibility for good accuracies along the null axis makes this antenna useful for narrower bandwidth guidance systems. A second approach to simultaneous multibeam DF makes use of a multimode, multiarm spiral in conjunction with a corporate feed network to utilize more accurate processing techniques.

Multiple-arm spirals can be configured for any of the types of planar or conical antennas discussed. The concept is simple enough: Four or more spiral arms are wound from the center following the equations for antenna growth presented above. Each arm of the spiral is separately fed by a port of an excitation matrix designed to produce simultaneous excitation of both odd and even modes by the action of the sum and difference properties of a 180 degree hybrid. Practical considerations, however, have led to the common use of four- and six-arm spirals. Reasons for this are obvious: It is difficult to feed more arms at the center; conical spirals become limited in high-frequency response due to the added truncation of the cone to accommodate more feeds; planar spirals become less self-complementary and exhibit different impedances, and so on.

Much work has been done to try to build multiple-arm antennas. The literature abounds with attempts to feed the spirals from the outside diameter (14–16); however, unwanted bands of radiation develop, restricting the RF bandwidth and causing high VSWR variations since the outside feed impedances vary with frequency. For these reasons we shall confine our discussions to the two most popular multiarm spirals, the four-arm sum (Σ) and differences (Δ) feed and the six-arm amplitude comparison type. Both designs are true monopulse approaches to passive direction finding and find practical use in many applications.

When a four-arm spiral is excited by a dual-mode excitation network, a sum or principal mode will develop, peaking along the bore-sight axis in a manner similar to the two-arm spiral described above. (This is not to be confused with the balun-fed spiral, which, although developing the same pattern, does so directly as opposed to summing process used in this case.) The network that accomplishes this is shown in Figure 3-21. Three hybrids and a 90 degree phase shifter form the excitation network. A portion of all elements is progressively summed and identified as the sum (Σ) output. The first level of hybrids form the 180 degree phase difference of opposing arms 1–3 and 2–4, one of which is shifted 90 degrees and added to the other to form the difference pattern.

The angle off boresight is θ; the angle around the circumference of the antenna is ϕ. If a signal is rotated around ϕ about the principal beam around bore sight, the antenna will register a 360 degree variation as indicated in Figure 3-22, which is a view of the phase variation in the X–Y plane parallel to the flat plane of the antenna. [Note that this is precisely the same pattern as that of the two arm due excited spiral discussed previously (Fig. 3-20).] This variation is called the first mode that undergoes a 360 degree phase change for a 360 degree spatial change and corresponds to the sum pattern defined as

$$f(\phi) = e^{j\phi} \tag{3-9}$$

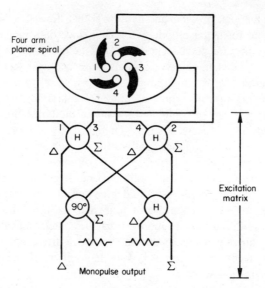

Figure 3-21. Four-arm spiral spatial monopulse direction finder.

The difference (Δ) or second mode pattern, however, undergoes a variation of 720 degrees for the same 360 degree spatial variation for the second mode circle, corresponding to

$$G = G(\phi)e^{j2\phi\Delta} \qquad (3\text{-}10)$$

The relative phase difference between the two modes for the four-arm case is 360 degrees, and measurement of this difference phase will yield ϕ. This is the same

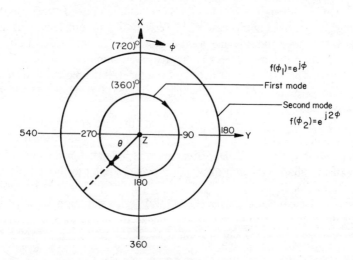

Figure 3-22. Phase relationships of the four-arm spiral.

as for the two-arm case previously discussed; Figures 3-20 and 3-22 are actually identical. The four-arm network is more predictable, however, and is the preferred technique.

The ratio of the sum and difference amplitudes will yield the angle θ; however, since it is the angle off bore sight between the two patterns in any direction, it will be ambiguous unless φ is used to supply a correction to derive the actual elevation angle. A processor designed to accomplish this is described in Chapter 4, Section 4.4.

One of the limitations of the two-arm spiral in multimode applications is the increase in axial ratio that occurs at far angles on either side of the boresight. The problem is exacerbated by the narrow (70) beamwidth of the peak pattern since the axial problem becomes worse as a percentage of beamwidth. One successful solution (17) has been to use a six-arm spiral to array the patterns, achieving a greater effective beamwidth per antenna of about 90 degrees. Figures 3-23 and 3-24 show the six-arm excitation matrix and the total processing function resulting in the development of four amplitude beams spaced 90 degrees from each other and corresponding to a four-quadrant DF system. Although this system appears complicated, it can be implemented using printed circuit technology resulting in a highly compact and production-repeatable design. Figure 3-25 shows the patterns resulting from this configuration, which are essentially 90 degree cardioids with acceptable axial ratios along the boresight. Deterioration of the axial ratio off boresight, however, still limits this approach to hemispherical measurements such as radar tail warning or missile guidance.

The processing for the above system can make use of a summed (Σ) output, which will be the approximate equivalent of the output of a simple spiral, as discussed. It is interesting to note that there will always be one less useful beam than there are arms in multiarm spiral configurations. This is due to the fact that $n - 1$ excitations are required for the noninfinite number of elements case. In the above six-arm spiral, only five useful patterns are available, for example.

3.2.6 Millimeter Wave Spirals—The Antector

Attempts to extend the high-frequency coverage of the spiral antenna into the millimeter frequency range have, up to the present, met with limited success. Although it is possible to get planar archimedian spirals to exhibit the familiar Gaussian patterns past 100 GHz, feed networks and especially the Marchand balun have not been successfully developed. Transverse electric and magnetic (TEM) mode coaxial cable exhibits high losses and develops extraneous modes of operation above 40 GHz, making this type of transmission line useless at higher frequencies. Waveguide feeds are limited in coverage at the low end of a frequency range by the guide cutoff frequency and at the high end by unusable modes. Attempts to increase the frequency of coverage of waveguides by ridging techniques are generally too lossy.

A more fundamental approach has been developed by the design of a combined antenna and detector (18), which has been named the antector, a euphemism that

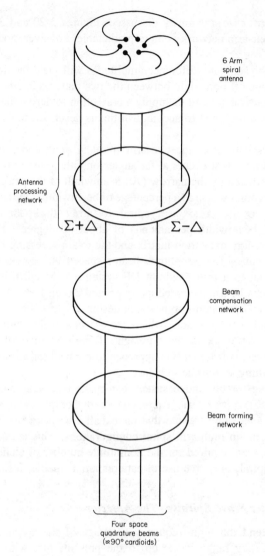

Figure 3-23. Excitation matrix for six-arm spiral antenna.

describes its operation. Basically, the idea consists of mounting a beam lead diode detector directly at the inner center feed point of a wideband MM wave spiral and coupling off the resulting detected video from one outer arm. This eliminates all of the problems of coaxial feed and all of the difficulties attendant to the use of many different suboctave band coverage waveguides. The other arm is used as a convenient point for application of direct current (DC) bias for the diode. Both arms are terminated to prevent RF reflections and decoupled to pass the bandwidth

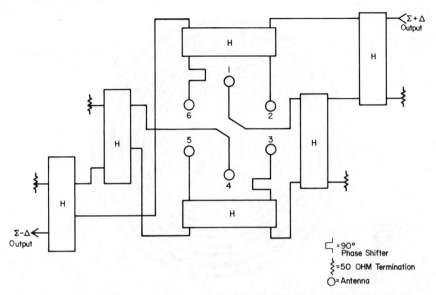

Figure 3-24. Antenna processing network for a six-arm spiral yielding $\Sigma + \Delta$ and $\Sigma - \Delta$ outputs.

of the millimeter wave detected video, which occupies a wide bandwidth associated with the narrower pulses found in this range. The concept is illustrated in Figure 3-26, which is a diagrammatic representation of the approach. The detector diode is chosen to present an impedance of approximately 100 ohms, and, since current can flow through it, acts like a balanced feed. The principal or odd mode of the spiral is excited causing the antenna to develop a beam that is peaked along the boresight of the axis perpendicular to the plane of the printed spiral elements. As for all wideband spirals, it is necessary to mount the radiating printed circuit in front of an anechoic absorbing material to cancel the backwave.

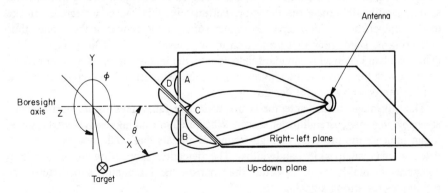

Figure 3-25. Forward-looking application of four-beam six-spiral antenna and feed network.

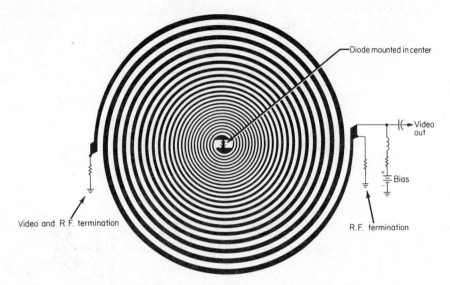

Figure 3-26. Diagrammatic representation of the antector showing electrical connections.

Figure 3-27 is a photograph of a beam lead diode mounted on a spiral that covers the frequency range from 18 to 100 GHz. Patterns have been taken at 20, 40, 65, and 100 GHz for the model photographed and are shown in Figures 3-28 to 3-31. A beamwidth of approximately 60 degrees is obtainable over this band with all of the characteristic properties of a spiral, low sidelobes, and a good axial ratio. Sensitivities in the order of lower band antennas and separate detectors can be achieved; however, the antector's sensitivity must be corrected for the diode's efficiency loss at MM wave frequencies.

There are two potential problems with the antector: possible burnout from high-power illuminators and an undesirably slow sensitivity rolloff at the low-frequency band-end limit. Since the antector is not shuttered or shielded, strong out-of-band signals readily illuminate the detector. Fortunately, this causes a decrease in the diode's impedance, which tends to reflect some of the power at the diode, thus protecting it. Data show that protection up to a level of +25 dBm is achievable. (This problem is solved in standard detectors by the incorporation of a limiter diode in shunt with the detector, a technique not feasible here.) The low-frequency rolloff can be improved by careful design.

The advantage of the antector is the wide and fairly predictable performance attainable in a low-cost millimeter sensor. When used in warning receiver applications, it can indicate the presence of high-frequency threats that could not be as readily detected over such wide bandwidths. The detection probability of this method compares favorably to or is better than narrow-band scanning superheterodyne techniques for strong signals.

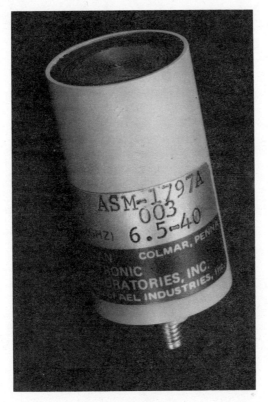

Figure 3-27. Antector: A combined cavity-backed spiral antenna and detector. Courtesy of AEL.

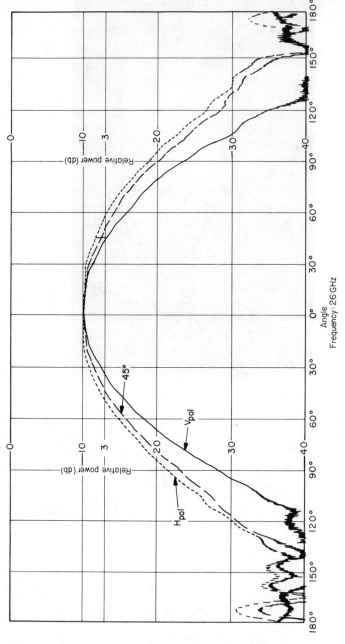

Figure 3-28. Azimuth pattern coverage of the antector at 26 GHz.

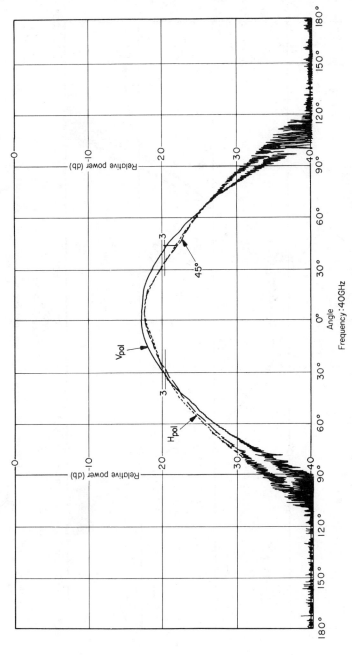

Figure 3-29. Azimuth pattern coverage of the antector at 40 GHz.

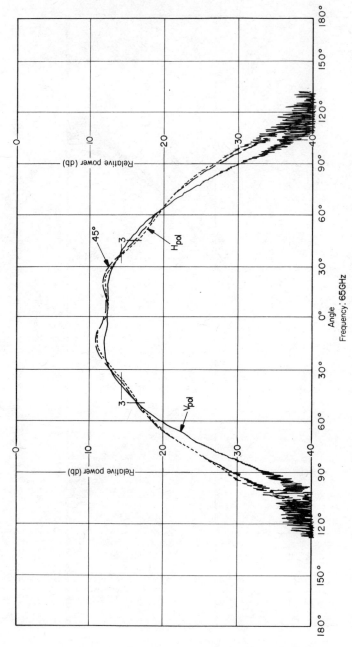

Figure 3-30. Azimuth pattern coverage of the antector at 65 GHz.

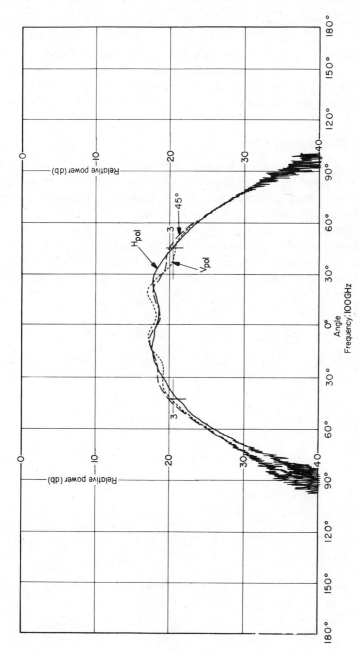

Figure 3-31. Azimuth pattern coverage of the antector at 100 GHz.

67

3.3 HORN ANTENNAS

A horn antenna is basically a transition from a transmission line to free space, taking into account the requisite impedance and field requirements. A brief description of transmission line theory will illustrate how this transition can take place for both coaxial and waveguide elements.

3.3.1 RF Transmission Lines

The following is an expository explanation of transmission lines in general and waveguides in particular. The explanations have been designed to be pertinent to the discussions on horns, biconical radiators, and polarizers and can be augmented by reference to many available texts (see Ref. 19).

Coaxial cable is perhaps the most useful type of RF transmission line since it exhibits a principal mode of propagation, called the transverse electric and magnetic, or TEM mode, capable of operating from DC to the high microwave frequencies. Figure 3-32 shows the construction of familiar coaxial cable, which consists of an

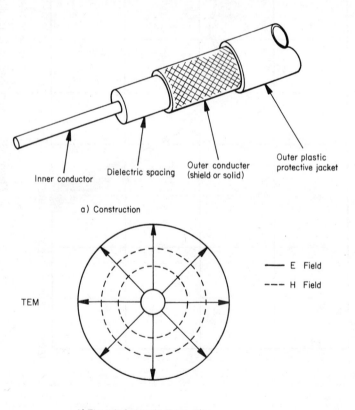

a) Construction

b) The principal mode of operation

Figure 3-32. Coaxial transmission line. (*a*) Construction. (*b*) Principal mode of operation.

inner conductor centered within an outer conductor, that may be braided for flexibility. A solid copper conductor is used for semirigid lines. The area between the two conductors is filled with a low-loss dielectric material, usually polyethylene or Teflon, which may be solid or made of beads. Flexible line has an additional outer polyvinyl chloride plastic protective coating, whereas semirigid line may be plated.

Energy propagates in the dielectric as the frequency increases. The E and H fields are traverse to the direction of energy propagation; that is, the E and H fields lie in a plane perpendicular to the signal flow. The two conductor sizes and the characteristics and separation of the dielectric material determine a "characteristic impedance," which is a physical property of and is used to identify the line. When terminated in its characteristic impedance, the line exhibits a resistive input impedance at the source end for a resistive termination at the far end.

Coaxial cable is limited, as the frequency increases, due to a gradual buildup of loss in the dielectric material needed to support the inner conductor. There is also additional ohmic loss in the two metallic conductors due to surface "skin effect" as the signal travels down the line. At very high frequencies, near 30 to 40 GHz, coaxial cable exhibits additional cylindrical waveguide modes, which impose an upper frequency boundary. Very small dimensioned cable is available to overcome this problem; however, the loss is high (typically greater than 0.5 dB/ft), making long cable connections impractical.

Waveguide is the common form of microwave transmission line used because it has low loss and is dimensionally suitable at microwave frequencies. The low-loss properties derive from the use of air or gas as the dielectric in a conducting hollow tube or guide. Propagation of RF energy occurs as described by wave theory. Poynting's theorem can be used to give the direction of the power flow or flux at any point in a dielectric medium, as a function of the direction of the electric and magnetic fields. If E and H are the values of the fields associated with an alternating current at any point, then the power P in the dielectric is

$$P = EH \sin \theta \qquad (3\text{-}11)$$

where

$$\theta = \text{the angle between } E \text{ and } H$$

or in vector notation,

$$\mathbf{P} = (\mathbf{E} \times \mathbf{H}) \qquad (3\text{-}12)$$

which is the cross product of the **E** and **H** vectors.

This formula follows the right-hand rule; that is, if the fingers of the right hand curl in the direction of **E** to **H**, the thumb points to the direction of power flow. From this it is clear that propagation or movement of power takes place. (This is also true for the coaxial cable.) If a conductor is placed near the propagating wave, any component of the electric field parallel to the conductor sees an equipotential conducting surface that acts as a short circuit reducing the E field rapidly to zero.

If any component of the H field flows perpendicular to the conductor it encounters a short circuit that causes it to fall rapidly to zero. In a hollow rectangular waveguide, which is bounded by conductors on all four sides, both E and H fields cannot be transverse with respect to the direction of propagation due to the above effects; therefore TEM waves do not exist in waveguides.

Two classes of waves do propagate in a waveguide, however—the transverse electric TE_{mn} and transverse magnetic TM_{mn}. (The British call these waves H and E.) Subscript m refers to the number of half-wave variations in the X direction, and subscript n refers to the number of field variations in the Y direction. (To prevent confusion and to follow current convention, the X direction will be considered parallel to b, and Y will be parallel to a as shown in Figure 3-33.) The TE_{mn} wave contains the electric field vector perpendicular to the direction of travel (transverse) with no net component of the H field in the direction of propagation. The TM_{mn} wave has the magnetic perpendicular (transverse) with no net component of the electric field in the direction of propagation. A waveguide therefore bounds or

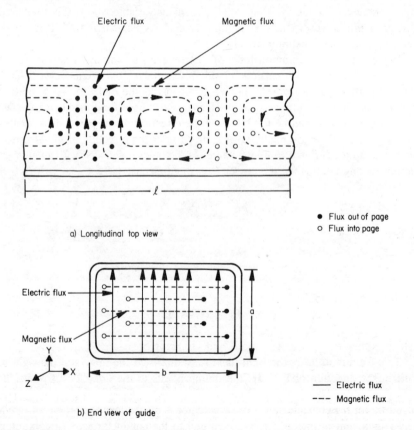

a) Longitudinal top view

• Flux out of page
○ Flux into page

Electric flux

Magnetic flux

——— Electric flux
– – – Magnetic flux

b) End view of guide

Figure 3-33. Representation of the TE_{10} (principal) mode of propagation in a rectangular waveguide. (*a*) Longitudinal top view; (*b*) end view of guide

contains the RF energy allowing only certain modes to exist due to the short-circuit effects of the conducting walls of the guide.

Of all the modes possible, one mode—the TE_{10}, or principal, mode shown in Figure 3-33—propagates at the lowest frequency. From Figure 10-33a, it may be seen that moving along the length of the guide, the electric field reverses its direction while the magnetic flux forms circular flows in alternating directions. The electric field crosses over (is zero) at the center of each magnetic flux circle. Note that the magnetic flux is bent when it is perpendicular to a wall due to absorption of the perpendicular wave, allowing only the magnetic flux parallel to the wall to flow.

From Figure 10-33b, we determine that the electric field is a maximum at the center of the guide (more lines) and diminishing to zero at the conducting boundaries. The value for $m = 1$ indicates that there is one-half wave of E in a cross-section transverse to the power flow along the width b. The $n = 0$ indicates that there is no variation of E along the a dimension.

The cutoff frequency is defined as the frequency of the TE_{10} mode or as the frequency of maximum or near-infinite attenuation. The lowest or principal mode occurs when

$$\lambda_c = 2b \sqrt{\mu\epsilon} \simeq 2b \qquad \text{for air} \qquad (3\text{-}13)$$

where

$$\mu = \text{the guide dielectric permeability}$$

$$\epsilon = \text{the dielectric constant}$$

This makes the waveguide essentially a high-pass filter, propagating all frequencies higher than λ_c where practical attenuation occurs. The useful operating range for most TE_{mn} guides is from 1.3 times the cutoff frequency to 0.9 times the cutoff frequency of the next higher mode. This results in about a 1.5:1 ratio for narrow height (height $\approx b/2$) guides. High-frequency propagation is limited by the appearance of higher-order TE and TM modes that limit simple waveguide to suboctave coverage. These properties are often made use of in the design of waveguide filters and couplers.

The height a determines the power-handling capacity of the waveguide since it must withstand the maximum E field strength. The ohmic loss of the guide occurs as the wave propagates along the length of the guide. The current of the H field perpendicular to the direction of wave propagation penetrates the wall, while the E field drops to zero at the conducting boundaries as described. The depth of penetration of the RF energy is

$$\delta = \frac{1}{\sqrt{\pi\mu f\sigma}} \qquad \text{in meters} \qquad (3\text{-}14)$$

Table 3-1. Wavelength Characteristics of Rectangular and Circular Waveguides

	Rectangular Guide[a]	Circular Guide[b]
λ_c(cutoff wavelength) (attenuation is infinite)	2.0 *b*	3.41 *r*
λ_g(usable guide wavelength)	1.6 *b*	3.20 *r*
λ_n(shortest wavelength before next mode)	1.1 *b*	2.8 *r*

[a] *b* is guide width.
[b] *r* is radius of circular guide.

where

$$\mu = 4\pi \times 10^{-7} \text{ for copper}$$

$$\sigma = 5.8 \times 10^{7} \text{ for copper}$$

$$f = \text{frequency in MHz} = 3000 \times 10'$$

For example, at $\lambda = 10$ cm, $\delta = 1.2 \times 10^{-4}$ cm, which shows that the penetrating fields and currents are at "skin" depth. Losses in a guide or coax operating at high frequency occur at the surface of the conductors making silver, rhodium plating or gold flashing popular techniques to reduce loss due to the high conductivity of those metals. Table 3-1 gives the relationships for the principal propagating modes of rectangular and circular guides for references.

The operating frequency of a waveguide can be made lower by adding one or two ridges as shown in Figure 3-34. The addition of the ridges reduces the power-handling capability of the waveguide, but for receiver and lower power transmitter applications it can be useful. The ridges can be continued into the horn for wideband applications. Losses in the guide using this approach are greater, as described in Ref. 20.

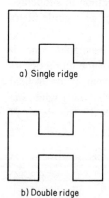

a) Single ridge

b) Double ridge   *Figure 3-34.* Ridge-loaded waveguide. (*a*) Single ridge; (*b*) double ridged.

3.3.2 Single-Mode Horn Antennas

Horn antennas find use in passive DF systems chiefly because of the gain improvement and pattern directivity they offer. In transmit applications, power-handling capability of the horn is important; however, in receiving applications, we are usually concerned with pattern ratios to form beams and predictable beamwidths both for direct signal measurement and for the purpose of illuminating passive reflectors that can shape and move the horn's beam as desired. A horn antenna is a transition structure that conducts a signal in a waveguide transmission line to an unbounded medium, such as air, where the signal is free to radiate. A waveguide has a characteristic impedance that must be matched to that of free space (120π ohms). Since, by Rayleigh's reciprocity theorem, an antenna is reciprocal (it works the same way when transmitting as when receiving); it will receive an inbound wave from free space and will transition it into a waveguide as well as carry an outbound wave from a transmitter to an antenna where it is radiated. The pattern and gain will be the same for either case.

In passive DF receivers, it is generally desirable to obtain maximum frequency bandwidth and control of the antenna's pattern gain and symmetry. Horn antennas, therefore, find application mainly in arrays where they are used to generate either a linear (in-line) or circular phase front with the mechanical advantage of close proximity "stacking." This is shown in Figure 3-35 for a lens-fed array where a set of closely spaced waveguides radiate, each acting directly as an antenna to

Figure 3-35. Waveguide array. Courtesy of Raytheon Corp.

generate a linear wave front. In rotating passive reflector systems, horns are used with rotary joints to illuminate a flat or focused reflector in a prescribed way to obtain specific patterns.

Gain is provided by the aperture size of a horn if the aperture fields are uniform in phase and amplitude. In this case, the beamwidth (BW) ϕ_{az} in either the x or y direction θ_{el}, as shown in Figure 3-36 is simply

$$BW_\phi = \frac{\sqrt{3}}{2}\frac{\lambda}{d_x} \tag{3-15}$$

$$BW_\theta = \frac{\sqrt{3}}{2}\frac{\lambda}{d_y} \tag{3-16}$$

The highest gain of the antenna is

$$G = \frac{4\pi}{\lambda^2}A \tag{3-17}$$

where

G = the uniform aperture gain

A = the physical area $(d_x d_y)$ of the aperture

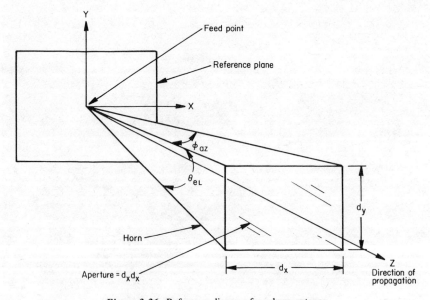

Figure 3-36. Reference diagram for a horn antenna.

Realizable gain, however, will be considerably lower due to ohmic and reflection losses. Practical numbers allow for 50% efficiencies. A useful relationship for gain as a function of the beamwidth for most EW applications where direct feed is used is

$$G = \frac{32,000}{\phi\theta} \tag{3-18}$$

where

$$\phi = \text{azimuthal beamwidth (degrees)}$$

$$\theta = \text{elevation beamwidth (degrees)}$$

A relationship may be found for reflector feeds in Ref. 10, where gains using 26,000 for the numerator in Equation (3–18) are good approximations for the illuminating efficiency used.

The minimum sidelobe levels of a constant illumination pattern is of the six x/x form which places the first sidelobe at -13.3 dB below the main beam. This is not much dynamic range for direction finding; therefore, to provide lower side-lobes, nonuniform illumination of the aperture obtained by varying the E field in a cos or $\cos^2$ manner is often used. Table 3-2 lists typical sidelobes levels for different illuminations.

Variation of the illumination can be seen to make substantial differences in the back- and sidelobe levels. These factors are vital in radar systems where reflector illumination is important; however, for passive DF systems using multiple direct radiating antennas, it is often possible to use the simple type of uniformly illuminated open-ended waveguide.

There are three basic types of rectangular waveguide fed sectorial horns: the E plane, the H plane, and the pyramidal as shown in Figure 3-37. The E plane gives a narrow elevation and wide or fan azimuth beam for application in height-finder radar systems and for sidelobe inhibition. The H plane horn gives a narrow azimuth and wide fan beam elevation pattern and is most popular for passive DF applications when only azimuthal angle of arrival is required, since maximum intercept probability

Table 3-2. Illumination vs. Side Lobe Level

Type of Illumination	Illustration	Side-Lobe Level (dB)
Uniform		-13.3
11.4 dB taper ($r = .862$)		-19.5
Cos		-23.1
$\cos^2$		-31.7

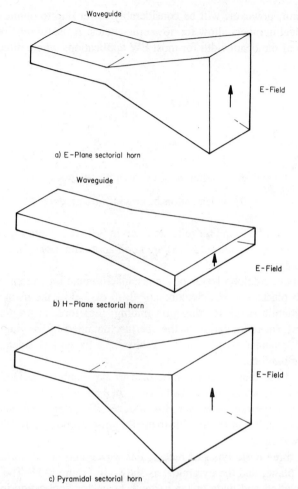

Figure 3-37. Types of rectangular horn antennas: (a) E-plane sectorial horn. (b) H-plane sectorial horn. (c) Pyramidal sectorial horn.

will be obtained in the vertical plane due to the wide elevation coverage. The pyramidal horn can provide a symmetrical beam ($\phi = \theta$). This is desirable for high gain narrow pencil beam raster scan systems where field of view is traded for gain, especially at millimeter wave frequencies where RF preamplification and low noise figures are difficult to obtain.

The polarization of a horn is determined by the direction of the E field vector, which is initially established by the waveguide feed. The polarization can be shifted to slant linear (E field at 45 degrees) by means of an internal or external polarizer, which converts the direction of the arriving E field to that of the antenna at the expense of some gain. Waveguides can also be twisted 45 degrees and fed to a pyramidal horn for slant linear polarization, which is a compromise. Various other methods to make horns respond to many linear and circular polarizations are used,

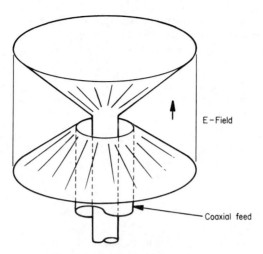

Figure 3-38. Biconical horn radiator.

especially in passive DF systems that deal with unknown signals. Some of these methods involve waveguide feed techniques in combination with hybrids.

The biconical horn shown in Figures 3-38 and 3-39 is another type of radiator that finds use both as an omniazimuthal antenna and as a DF antenna. This type of antenna, fed with a coaxial cable, supports both the transverse E and H field modes (TEM) and can be caused to support higher-order modes that can be used for direction finding. The polarization of the biconical feed, as shown, is vertical, although horizontal polarization can be obtained by other feed methods (21).

Figure 3-39. Honey–Jones biconical antenna fed by an overmoded coaxial structure. Courtesy of SRI International.

Horn antennas are not power limited for passive DF systems since maximum signal levels are seldom over +10 dBm. Direct crystal detection can be a problem in a horn/detector receiver since a high-power radar may illuminate and burn out the detector. Limiter diodes are often used ahead of the detector to prevent this. In radar work the horn designer must consider power-handling capacity and the possibility of arc breakdown, especially for high peak-power pulse transmitters. This often leads to increased height waveguide and the need for pressurization or use of an inert gas, which is costly since the radiating system must be sealed. An important passive DF problem is bandwidth. Since the waveguide horn only propagates energy at specific modes (the dominant being the TE_{10}), no propagation will take place below the cutoff frequency. This makes the waveguide act as a high-pass filter for the horn placing a practical low end limit on frequency coverage. High-frequency performance is determined by the appearance of other propagation modes in the horn and waveguide feed. Most are not usable and therefore essentially limit performance of the single horn to less than an octave of frequency coverage (see Table 3-1). It is possible to extend horn frequency coverage as in waveguide by adding a single or double ridge at the center. With proper design and by transitioning the ridge coverage of 10:1 can be attained (11). Many techniques have been developed to feed the ridges to obtain response to all polarizations or to make a linearly polarized horn exhibit circular polarization by using the phase angle changes introduced by tapering the ridges. Figure 3-40 shows a 9–18 GHz 160 degree azimuth by 60 degree elevation horn using double-ridged (WRD 750) waveguide to obtain the wide bandwidth. An external polarizer is used.

Figure 3-40. An example of an octave band (9–18 GHZ) Horn Antenna using double ridged waveguide.

3.4 MULTIMODE DF HORN ANTENNAS

A single horn antenna can be made to exhibit properties of multiple horns to develop monopulse patterns by making use of various mode characteristics of waveguides. An interesting example of this can be found in a single aperture multimode direction-finding system described in Ref. 22. Consider the phase monopulse system described in Chapter 2: Two parallel antennas, closely spaced, receive signals on axis equally; hybrid subtraction is used to develop a null on boresight. A signal displaced off axis in either direction will develop a phase (or amplitude depending on the antenna orientation) difference signal that is proportional to the angle on either side of bore sight. In radar, the antennas for angular determination are physically constructed and mounted in a cluster of four to provide simultaneous azimuth, elevation, and sum patterns. Howard and Lewis of NRL (ref. 18, 19, and 22) recognized that the same outputs from the cluster could be obtained by exciting multiple waveguide modes simultaneously in a single aperture.

3.4.1 A Single Aperture Multimode DF Horn

In Figure 3-41, the TE_{10} mode in a rectangular wavelength is shown as a sine wave or symmetrical field with a maximum at $b/2$ and 0 at each equipotential boundary (per our convention). The TE_{20} mode of this guide when excited by an off-bore-sight axis signal will develop a reversal of the E field at $b/2$ and will exhibit sinusoidal field properties. In effect, the TE_{20} mode is an asymmetrical version of the TE_{10} mode that satisfies the second postulate of monopulse regarding skew symmetry. In actuality, any signal received by a TE_{20} mode waveguide horn will only excite

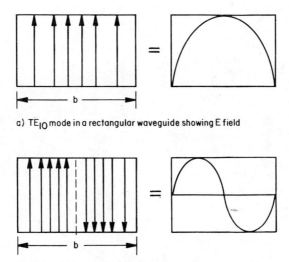

a) TE_{10} mode in a rectangular waveguide showing E field

b) TE_{20} mode in a rectangular waveguide showing reverse of E field

Figure 3-41. Mode relationships showing (*a*) TE_{10} and (*b*) TE_{20} field distribution for direction finding.

a nonzero output if it is off boresight and if each half of the mode is detected and subtracted. The relationships between the modes are directly analogous to the pattern relationships required for monopulse systems. This was recognized by Howard (22) who constructed and described many variants of feeds using this principle. An azimuth-elevation monopulse feed was built using the TE_{10} mode as a reference signal (Σ), the TE_{20} mode as the H plane (Δ_θ), and a combination of the TE_{11} and TM_{11} modes to give the E plane difference (Δ_ϕ). The structure was built in a square waveguide ($E = H$)) to maintain the phase relationships of the modes. These mode relationships were transitioned into a single aperture horn antenna and used to provide the standard monopulse feed network.

The advantage of the concept of mode-derived patterns is that excellent null stability is attained since accuracy depends upon the properties of the mode excitation and support as a first-order effect as compared to factors in the construction, shape stability, and physical properties of direct radiators. Another advantage is the derivation of the two horns' monopulse direction-finding properties from a simple radiating feed horn structure as compared to trying to match two different horns. Consider a simplistic single plane (azimuth) version of a single-axis mode-derived horn

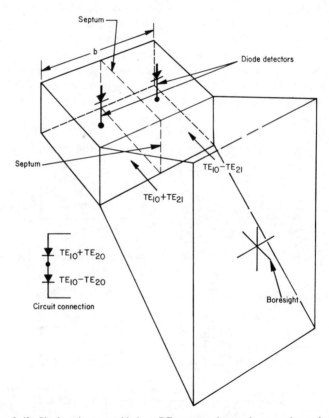

Figure 3-42. Single axis waveguide horn DF system using mode separation techniques.

antenna system as shown in Figure 3-42. In this horn of width b, the TE_{10} mode is dominant; however, the TE_{20} mode can be excited by probes or antennas placed within the guide. This is accomplished by the two detector diodes that are located where the TE_{10} and TE_{20} signals are both present. A septum (conducting wall) can be placed between the diodes since the TE_{20} mode is zero for a boresight signal. The configuration is analogous to two-phase monopulse antennas placed side by side. Each diode will see a $TE_{10} + TE_{20}$ sum, however, since the TE_{20} field reverses at the other side of boresight; a series connection of the diodes as shown will give a positive output for a signal at one side of boresight and a negative signal for a signal at the other side, with zero output for a signal directly on boresight. It is assumed the horn must support all the modes. The septum is a matching device to assure proper phase relationships at the diodes. A more complex arrangement can be used to provide higher sensitivity and both azimuth and elevation patterns by utilizing hybrids to isolate each of the modes, which are then detected and compared by superheterodyne methods. The simple system described here can only operate over a relatively narrow (suboctave) bandwidth, which is a limitation that can be overcome with more modern technology utilizing strip-line and microstrip techniques. The example shown here illustrates mode filtering as a method of beamforming. This technique is not limited to rectangular or square waveguides.

Spiral antenna methods are more generally used than mode-formed horn techniques since they offer wider RF bandwidths. Horn methods are useful in arrays or whenever narrower beamwidth patterns can be utilized to give more gain. Polarization is a second problem in both single- and multimode horns since in passive direction finding the input polarization is not likely to equal that of the horn. This favors the use of circular polarization where the spiral has a definite advantage. Slant linear polarization is a useful compromise solution though, and horn systems can be twisted 45 degrees or can make use of internal or external polarizers to respond accordingly.

3.4.2 The Biconical Multimode (Honey–Jones) Antenna

The biconical horn radiator is capable of supporting the TEM as its principal mode, confining all of the TEM energy in the field between the two bicones. The coverage of this mode of operation is omniazimuthal ($\phi = 360$ degrees) and nondirectional. This may be deduced from the TEM mode field for the coaxial feed. (An exploring dipole rotated around the dielectric of a coaxial cable at a fixed radius r would measure a constant or omniazimuthal E field.) The advantage of this type of antenna is the wide frequency range of coverage, which can be almost 20:1.

The concept of using the biconical antenna to perform direction finding was first recognized and described by Honey and Jones (23) in a paper describing a biconical antenna fed by a coaxial to waveguide structure capable of supporting and separating the TEM omniazimuthal and TE_{10} directional modes of the biconical antenna. The TEM mode is omniazimuthal and constant in amplitude as a function of angle of arrival. The TE_{10} mode has a sinusoidal amplitude variation proportional to the angle of arrival, which rotates at a uniform angular velocity as a function of the RF frequency. All the DF information of the system is contained in TE_{10} mode,

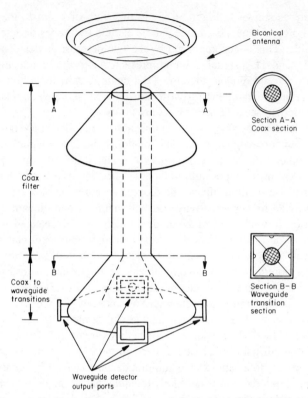

Figure 3-43. Mechanical configuration of the biconical direction finder.

either as an amplitude that varies with input angle or as an RF phase difference between it and the TEM mode. Two methods of extraction of the DF angle were described: The first was a wide instantaneous video bandwidth video detection system; the second a mode separation phase measurement superheterodyne technique.

The original wideband video method used a coax-to-waveguide transition that converted the TEM and TE_{10} antenna modes into TEM and TE_{11} coaxial modes, which were then further transitioned into four TE_{10} ridged waveguide outputs, each representing a quarter, or 90 degree, view of the coaxial structure. Square law detectors placed at each waveguide port took the sum of the omniazimuthal and TE_{11} components, detected them to remove phase differences, and then performed opposite pair subtraction to provide the resulting output video. This gave an output similar to that of four orthogonally placed spiral antenna/detector combinations, except that the biconical beamwidths were mathematically 90 degrees wide and independent of frequency, an optimum condition for a balanced system. Figure 3-43 shows the essentials of the feed: A biconical antenna is fed by an overmoded uniform diameter coaxial structure capable of supporting the TEM and TE_{11} modes.

An RF signal at a frequency where both modes are supported, arriving at an angle ϕ, establishes the TEM and the TE_{10} modes in the biconical antenna at that angle. These modes transition into the TEM and TE_{11} modes in the uniform length of coaxial line capable of sustaining them. At section B–B, a transition of each quarter of the coaxial cable into four TE_{10} waveguide outputs is gradually made. Figure 3-44 shows the mode relationships along the signal path. The TEM mode moves rapidly from the antenna to the bottom of the coaxial structure, alternating in field direction. The TE_{10} mode of the antenna is bent and initiates the TE_{11} mode at the input to the coaxial line at an angle proportional to ϕ. The resultant energy travels in two components that are orthogonal (of the form of the sine and cosine of ϕ). Propagation of this mode is spiral (like a circularly polarized wave) and is slower than the TEM mode, arriving at the bottom of the feed at a later time at an angle δ, representing the phase (time) difference between the TEM and TE_{11} modes. The maximum magnitude of the TE_{11} mode is directly proportional to the input angle ϕ rotated by as many degrees as required to travel the length of the coaxial line or "mode filter."

In the wideband DF system described by Honey–Jones, four VSWR-matched detectors were placed at each port providing outputs. The predetection signals are as follows:

$$E_1 = A \sin \omega t + B \sin \phi \sin(\omega t - \delta) \qquad (3\text{-}19)$$

$$E_2 = A \sin \omega t + B \sin \phi \sin(\omega t - \delta) \qquad (3\text{-}20)$$

$$E_3 = A \sin \omega t + B \sin \phi \sin(\omega t - \delta) \qquad (3\text{-}21)$$

$$E_4 = A \sin \omega t + B \sin \phi \sin(\omega t - \delta) \qquad (3\text{-}22)$$

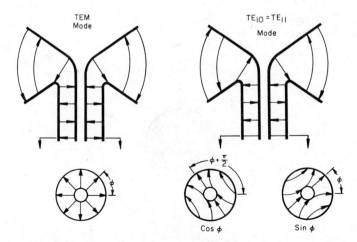

Figure 3-44. Mode relationships in multimode biconical antenna and coaxial feed.

where

A sin ω*t* contains   ¼ of the TEM power

B sin φ and B cos φ = the two direction-carrying TE_{11} mode signals

sin(ω*t* − δ)   = the term containing δ, the phase lag between the TEM
                                             and TE_{11} modes

Detecting these signals provides four video outputs:

$$E_{1D} = A^2 + 2AB \sin \phi \cos \delta + B^2 \sin^2 \phi \qquad (3\text{-}23)$$

$$E_{2D} = A^2 + 2AB \sin \phi \cos \delta + B^2 \sin^2 \phi \qquad (3\text{-}24)$$

$$E_{3D} = A^2 + 2AB \sin \phi \cos \delta + B^2 \sin^2 \phi \qquad (3\text{-}25)$$

$$E_{4D} = A^2 + 2AB \sin \phi \cos \delta + B^2 \sin^2 \phi \qquad (3\text{-}26)$$

An input signal will have twice its power in the TE_{11} mode as the TEM mode, hence $B = 2A$.

Examination of the detected equation set shows that each signal consists of an angle-of-arrival term (the middle), which contains the bearing information φ that varies in output as a function of the phase lag (maximum output occurs for δ = 0, the no lag case). By subtracting opposite detector pairs, sin φ cos δ and cos φ cos δ terms containing all angle information unambiguously result. The subtraction process is necessary to develop unambiguous data, since the presence of the cos δ creates a backlobe, which can equal the frontlobe when δ = π/2. Under these conditions, the subtraction will cause the outputs to fall to zero. Cos δ, then, limits the total bandwidth of the antenna by defining the operating range to be between a fixed limit of output, 2 dB being chosen in the original reference. The coaxial mode filter is limited at the high-frequency end by the presence of a higher order mode, the TE_{21} shown in Figure 3-45. This mode identifies itself as a variation appearing every 90 degrees.

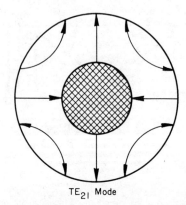

TE$_{21}$ Mode

Figure 3-45. The $TE_{2,1}$ higher order mode in the overmoded coaxial feed.

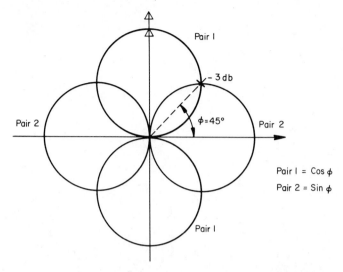

Figure 3-46. Subtracted output pairs of the ideal multimode biconical direction finder.

The original antenna covered 8.3–12.4 GHz with peak errors ranging from ± 3 to ± 7 degrees at the high-frequency end. The output patterns were sinusoidal in field as shown in Figure 3-46 and exhibited sharp DF video nulls of 21 dB. The limitations of this approach were clear; however, a long coaxial section desirable to attenuate the TE_{21} mode at the high-frequency end exacerbated the variation of the phase shift δ causing the antenna output to fall. Detector matching and unequal sensitivities created additional errors. These problems led to the concept of separation of the TEM, TE_{11} sine, and TE_{11} cosine modes through the use of quadrature couplers and magic "T" hybrids. This technique led to several methods for DF determination that were unambiguous when using narrowband superheterodyne techniques to resolve phase ambiguities. The wideband methods were, unfortunately, ambiguous due to the inability to resolve the subtracted sine and cosine 180 degrees ambiguity.

Later developments using the Honey–Jones antenna continued. An L-band superheterodyne direction finder utilizing the mode-separation filter technique was developed by Chubb, Grindon, and Venters (24). This system, shown in Figure 3-47, utilized a turnstile junction and a series of hybrids, as shown, to form three output voltages:

$$V_1 = 4A \cos \omega t \tag{3-27}$$

which is the TEM omniazimuthal signal

$$V_2 = 2B \cos(\omega t + \phi) \tag{3-28}$$

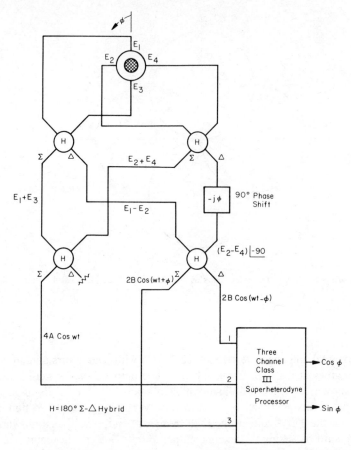

Figure 3-47. Narrow band superheterodyne phase monopulse direction finder (after Ref. 24, Chubb et al.).

the TE_{11} right-hand circularly polarized component and

$$V_3 = 2B \cos(\omega t - \phi) \tag{3-29}$$

the TE_{11} left-hand circularly polarized component.

To process these signals, a modified version of a three-channel class III processor was used, with an additional 90 degree preprocessing phase shift added.

By examination, it can be seen that only the V_1 and either of V_2 or V_3 need be used to obtain ϕ; however, the system was used to detect moving aircraft, which create multipath interferences that continually change. By measuring ϕ using V_2 and V_3 and averaging the results, the multipath component can be made to cancel. The resultant improvement in accuracy was substantial (about 6:1), with a final error of approximately 5 degrees RMS. The signals V_1, V_2, and V_3 were phased compared to V_1 to determine ϕ. The phase dispersion δ between the TEM and TE_{11} modes must be subtracted out, which required a knowledge of frequency that was not difficult here since detection was at a known beacon frequency. The TE_{21} problem was not encountered since operation was below the frequency at which this mode appeared. Although partially suited to the objectives of wideband passive direction finding, the above system worked with cooperating aircraft as part of a collision avoidance technique to demonstrate feasibility. For wide RF bandwidth, a scanning superheterodyne receiver could give frequency information for postdetection δ compensation that would permit the receiver to scan over a suboctave range.

A wider RF bandwidth antenna system had been envisioned by Honey but was not achieved until two important facts were realized: Mode separation and cancellation of the TE_{21} mode could be achieved by proper design of a hybrid feed network and the phase delay between the coaxial TEM mode and the waveguide TE_{11} modes could be compensated for prior to detection. In addition, better methods could be developed to extract the various modes by utilizing turnstile RF mode techniques as undertaken by Chubb et al. (24).

The Honey–Jones antenna has always been attractive for direction finding because of the symmetry of its design and the natural circular phase front exhibited by the bicone. It is also possible to provide an elevation (H plane) tilt for above the horizon applications to minimize ground clutter. This better control of elevation is a prominent reason for its continuing development.

The original Honey–Jones antenna was a four-port device; that is, there were four RF outputs, each containing the sum of the TEM and TE_{11} modes. These were developed by a transition of the coaxial feed to waveguide. A later approach considered the use of five ports. The first was a standard coaxial output mounted at the bottom of the feed to extract the TEM or omniazimuthal mode directly. The others consisted of four orthogonally spaced coaxial probes whose isolated center conductor intercepted the dielectric of the coaxial feed network, much as an antenna would. These four probes lay in a plane perpendicular to the direction of the feed and were spaced 90 degrees around the circumference. Figure 3-48 is a diagram of the arrangement. The five-port device effectively formed a turnstile junction with the advantage of a separate omniazimuthal output. This permitted a wide open azimuthal coverage for other purposes such as instantaneous frequency measurement, while the four DF ports could be used for direction-of-arrival determination on a pulse-by-pulse basis.

Although the above described antenna was built and fulfilled its design objectives, it became evident that the separation of the modes indirectly, by a hybrid excitation matrix, could provide the TEM output and offer additional advantages. Since this is perhaps the most significant configuration of the Honey–Jones concept, it will be explained in more detail. Figure 3-49 shows the approach. Four TE_{11} probes receive the energy that has propagated down the coaxial filter connected to the multimode bicone antenna. Each probe will detect the sum of the TEM component, the TE_{11} value, and the undesired TE_{21} mode, assuming operation for the worst case frequency where all modes occur. There will also be a phase angle δ representing

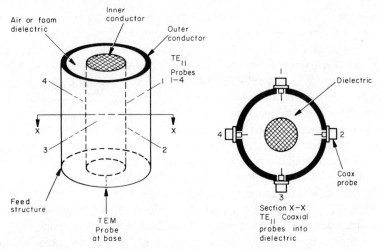

Figure 3-48. Five-port turnstile biconical antenna feed with mode separation.

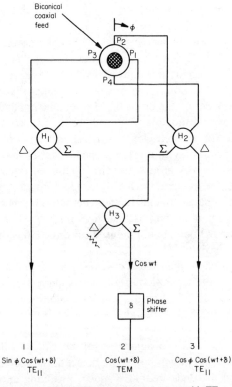

Figure 3-49. Three-port wideband Honey antenna system with TE_{21} mode canceled.

the delay in propagation of the TE_{11} and TEM modes. Hybrids 1 and 2 are connected to alternate probes and provide the sum and the difference output of the signals. This is done at a loss of 3 dB in power or a .707 ($\sqrt{2}/2$) in voltage since 3 dB hybrids are used. The sum (Σ) outputs of H_1 and H_2 are combined in H_3 and appear at the sum (Σ) port of H_3 as a cos ωt term. This output is delayed in a phase shifter with a phase angle δ to compensate for the faster propagation of this mode down the overmoded coaxial filters. (The phase shifter is actually a piece of waveguide exhibiting both the delay and frequency dispersion characteristics of the antenna feed structure.) The final output at port 2 is of the form cos ($\omega t + \delta$), which is seen to be the omniazimuthal TEM mode of the antenna. The proof of this is complex; a simplified method will be used for understandability.

Consider Figure 3-50, which shows the polarity of a voltage of unity induced in each of the probes for each of the modes of propagation, the TEM, TE_{11}, sine, TE_{11} cosine, and TE_{21}. The convention used here is that the field will induce a negative polarity when it is in the same direction as the probe and will induce no voltage (shown for the TE_{11} case) when it is orthogonal or perpendicular to the probe. Since the antenna is a linear device, the law of superposition holds and each mode can be analyzed separately. Since the fields are symmetrical, it will be assumed that equal voltages are induced at opposite ports. Reexamining Figure 3-49 for the TEM mode, probes P_3 and P_1 are fed to H_1, and P_2 and P_4 are fed to H_2. The

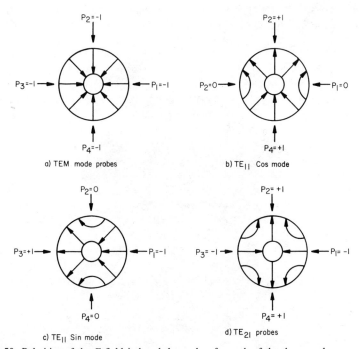

Figure 3-50. Polarities of the E field induced the probes for each of the three modes present at the bottom of the coaxial feed (E pointed IN is negative). (a) TEM mode probes. (b) TE_{11} cos mode. (c) TE_{11} sin mode. (d) TE_{21} probes.

polarity of the induced voltages in each probe is -1; therefore at the difference ports of each hybrid, there will be 0 TEM V and at each sum port, there will be -1.414 ($2X - \sqrt{2}/2$ of each input) V. At hybrid H_3, the difference port will have 0 TEM V, and the sum port will have the sum of all four ports or -2.28 ($\sqrt{2}/2$) V. The result is that hybrid H_1 has separated out the cosine component of the TE_{11} mode. By similar reasoning, hybrid H_2 separates out the sine component of the TE_{11} mode. The final output voltage of each mode is shown at port 1 to be $\sin \phi \cos(\omega t + \delta)$ and at port 3 to equal $\cos \phi \cos(\omega t + \delta)$.

The TE_{21} mode analysis is interesting: At the Δ port of H_1 the TE_{21} mode cancels, and at the Σ port -1.414 TE_{21} V is present. At H_2, cancellation again occurs for the TE_{21} mode, at the Δ port leaving $+1.414$ TE_{21} V at the Σ port. When the two TE_{21} V are combined in H_3, they cancel and are terminated in the load placed at the Δ port. All TE_{21} energy has therefore been removed by the feed matrix, and the effects of the TE_{21} mode disappear. The Σ port of H_3 contains only the TEM signal. This is a significant consequence since the antenna can now operate beyond the previously limiting upper frequency. The second limitation has been already minimized by the addition of the δ filter in the TEM mode; comparison of the three outputs are now all in phase. The result of this feed method permits the antenna to achieve full octave coverage by connection to a superheterodyne three-channel class II processor (logarithmic amplifiers with detectors and subtractors) to form the $\tan \phi$ or $\cot \phi$ amplitude ratio. This removes the $\cos(\omega t + \delta)$ term in the detection process. By noting if sine is greater than cosine, division by zero can be avoided and the appropriate tangent or cotangent output is used. Figure 3-51 is a

Figure 3-51. Three-output Honey–Jones antenna and polarizer. Courtesy of SRI International.

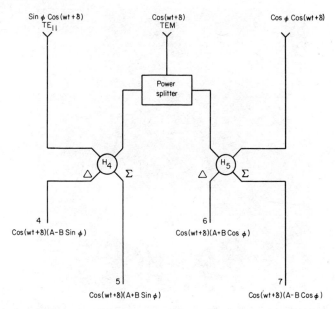

Figure 3-52. Recombination network for four-quadrant square law detector output.

version of the antenna that has the three outputs described above, as was used with the class II processor. Since the biconical antenna is vertically polarized, the figure shows it mounted within a "meander line" polarizer permitting reception of horizontal, vertical, and circular polarization.

The direction finder described above, although offering wide RF bandwidth, requires the class II processor, which is a superheterodyne receiver with logarithmic IF amplifiers. This processor provides linear detection due to the high-gain, high-level detection process and, invariably, is a relatively narrow band, requiring a frequency scanning process to explore a frequency range.

The original concept of a broadband square law detected system can be realized by expansion of the three-hybrid network back into a four-output network. This may be done by recombination of the three modes with three additional hybrids as shown in Figure 3-52. The omniazimuthal signal, devoid of its phase difference, is split and fed to hybrids H_4 and H_5. Each hybrid also received either the TE_{11} sine or cosine signal. By analogous reasoning it may be shown that four outputs are obtained:

$$\cos(\omega t + \delta)(A - B \sin \phi) \tag{3-30}$$

$$\cos(\omega t + \delta)(A + B \sin \phi) \tag{3-31}$$

$$\cos(\omega t + \delta)(A - B \cos \phi) \tag{3-32}$$

$$\cos(\omega t + \delta)(A + B \cos \phi) \tag{3-33}$$

Figure 3-53. Broadband Honey antenna with multiport strip-line feed network. Courtesy of SRI International.

The TE$_{21}$ mode is canceled out by elimination in the Δ port of hybrid H_3 in Figure 3-49. The output of the system still is multiplied by cos δ, but the *dispersion* with frequency has been reduced to the extent that the equalizer can replicate the overmoded coaxial line dispersion. This is important since, in linear detector systems, the loss due to the dispersion affects ambiguity resolution. In quadratric systems, reductions of the dispersion permits a wider bandwidth for a given δ. In general, δ is nonzero in the overmoded line since a finite length must be used, which must have some value since it is not feasible generally to make it zero.

The structure of a complete hybrid-equalized quadratic detected antenna system is shown in Figure 3-53, a photograph of the bottom of the antenna. The feed network utilizes wide bandwidth strip-line couplers to provide equal power splits with excellent phase tracking. The biconical antenna is filled with a polyfoam structure to support a mode suppressor to eliminate any spurious horizontal response.

3.5 ROTATING REFLECTOR ANTENNAS

The rotating passive reflector type of antenna still finds wide application in DF systems due to its simplicity and potential for reasonable gain with good backlobe response. The effective aperture of a rotating reflector, however, is never that of the reflector area alone since there is an illumination efficiency associated with the process. This efficiency is dependent upon the shape of the taper of the beam of the illuminating antenna and the reflectivity.

An antenna such as a spiral or a horn can be rotated directly by feeding it through a mechanical rotary joint. This is sometimes done; however, it is not always possible to achieve the front-to-back-lobe ratios of a reflector since two additional degrees of control are provided: the shaping or focusing of the passive reflector to increase gain and the ability to adjust the illumination or taper of the radiating structure to optimize the beam shape and back-lobe response. Rotating antennas, however, are still used. Multiple-channel rotating DF systems have been built using sum (Σ) and difference (Δ) channels in multiple-channel rotary joints, often in narrow RF bandwidth radar applications. For wide RF bandwidth systems, as required in passive DF applications, however, direct rotating radiators are generally limited.

There are many approaches to rotating reflector antennas. If a simple flat plate passive reflector is rotated at an angle above a linearly polarized feed, the polarization will rotate with the reflector and become a function of pointing angle. For this reason, most rotating reflectors use circularly polarized feeds or rotate the antenna and the reflector as a unit, using an RF rotary joint to make connection to the antenna. This is possible at frequencies where a good rotating antenna can be built. Since most rotary joints are coaxial, it is necessary, for mechanical reasons, to go to a TEM or coaxial type of transmission line at the joint assembly. For most applications, this limits the frequency range to bands where coaxial line can be used, placing the upper limit at approximately 20 GHz.

It has almost always been an object in the design of a rotating reflector system to shape the resultant pattern so that it can exhibit fan beam characteristics. Another object at high frequencies has been to shape both the vertical and horizontal beamwidths into a narrow "pencil beam" for high gain, especially at very high frequencies (above 20 GHz), to attempt to overcome transmission path propagation losses. The smaller size of the reflectors at these higher frequencies is a tempting prospect; however, the need for a circularly polarized feed imposes a practical limitation. We shall consider both problems and cite examples of successful implementations of both techniques.

3.5.1 A Broadband Shaped Beam Rotating DF System

The early concepts for the design of a shaped reflector system go back as far as 1946, when Stavis and Dorne described horn illuminated elliptical paraboloid reflectors (25). The conclusions reached are shown in Figure 3-54. A reflector shaped elliptically in elevation will yield a fan or broad-beam pattern (usually shaped as a cosecant^2 H-plane pattern); a parabolic shape in azimuth will narrow the azimuthal beam. Other conclusions are that a circular polarized feed is required to provide constant illumination of the reflector and that the horn pattern must be symmetrical about the horn axis. Baffles were found experimentally to reduce the size of the horizontal reflector surface.

Work was undertaken at American Electronic Laboratories to develop a rotating DF antenna with a wide (2–18 GHz) RF bandwidth, a constant beamwidth, and a wide H plane (θ) pattern to permit predictable DF processing algorithms to be used for shipboard use. This work was described by Bohlman (26). Figure 3-55 shows

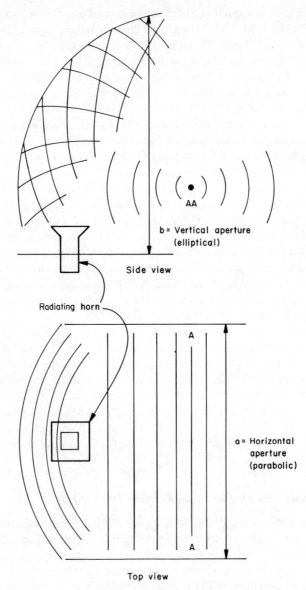

b = Vertical aperture
(elliptical)

Side view

Radiating horn

A

a = Horizontal
aperture
(parabolic)

A

Top view

Figure 3-54. Shape of a focused reflector for a narrow azimuthal (φ) and wide elevation (θ) beam.

Figure 3-55. A 2–18 GHz rotating DF system using a shaped passive reflector. Courtesy of AEL.

the design. A broadband 2–18 GHz circularly polarized spiral antenna is mounted at the common focus of a reflector designed to be parabolic in azimuth and elliptical in elevation. The spiral illuminates the reflector and is rotated with it through a 2–18 GHz rotary joint to eliminate pattern modulation due to the rotation and reduce squint. The reflector is 16 by 14 inches in size across these two curves. The ellipse that forms the azimuth beam has a second focus designed to minimize the aperture blockage due to the presence of the radiating spiral by placing it in front of the axis of rotation and 8 inches above the feed. Moving this focus point further out narrows the elevation beamwidth since the design of the elliptical section of the reflector will approach a parabola for widely separate foci. In the case of this design, a full fan beam was not desired in favor of increased gain. The reflector is constructed of laminated fiberglass sheets, honeycomb supported for added strength. The conducting surface is plated with a thin coating of silver coated with polyurethane to prevent oxidation. Absorber material is judiciously positioned to reduce reflectors. The bearing shaft of the reflector supports the rotary joint.

The feed antenna is a 2.3 inch circularly polarized cavity-backed archimedian spiral of fairly straightforward design. The radiation characteristic of this antenna is such that the beamwidth increases at the 2 GHz low frequency, which would tend to overilluminate the reflector, creating undesired backlobes. Since the reflector is parabolic in azimuth, this beam narrows; however, the elevation pattern in elevation increases. This may be seen in Figure 3-56a, which shows the azimuth beamwidth increasing to 23 degrees from a 5 degree average value over most of the band. The figure also shows the excellent axial ratio of this approach. Figure

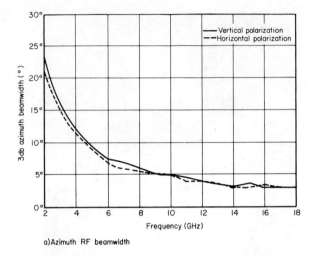

a) Azimuth RF beamwidth

b) Elevation angle measured at −6db down from beam peak

Figure 3-56. Rotating reflector antenna beamwidth. (*a*) Azimuth RF beamwidth. (*b*) Elevation angle measured at −6 dB down from beam peak.

3-56*b* shows the elevation for the 6 dB down off-peak beam gain, which increases at the lower-frequency end. The gain of the antenna and the boresight pointing error are shown in Figure 3-57*a* and *b*. Gain rolloff at the low end mirrors the increase in antenna beamwidth and efficiency of the spiral. The beam tilt is related to the sense of the circular polarization, the reflector size, and the feed offset (27, 28). A trade-off was made in this design between a fixed antenna mounted at the axis of rotation and the use of an offset feed and rotary joint. It was anticipated that the low-end beam variation would have caused a 10 degree beam tilt at 2 GHz with the fixed antenna, compared to the 3 degrees attained here. The final antenna, which is mounted in a full radome, sweeps a volume 22 inches in diameter by 16 inches in height. It provides an azimuth beamwidth of approximately 2.5–23 degrees over an *H*-plane 6 dB beamwidth of +20 degrees above to −5 degrees below the horizon. The front-to-back and all other lobe ratios are −20 dB down at all frequencies.

In another reflector-type antenna, the reflector was designed to provide a 1.5 × 2 degree pencil beam that could provide a narrow high scan. This required that the reflector be able to rotate in azimuth and scan in elevation. Figure 3-58 shows this antenna, which covers the 10–40 GHz range in three bands. The illuminating elements are three fixed circular polarized horns mounted as close to the axis of rotation as possible. The reflector rotates on a slip assembly that carries the circuit associated with the tilting mechanism. The reflector is a flame-sprayed conducting

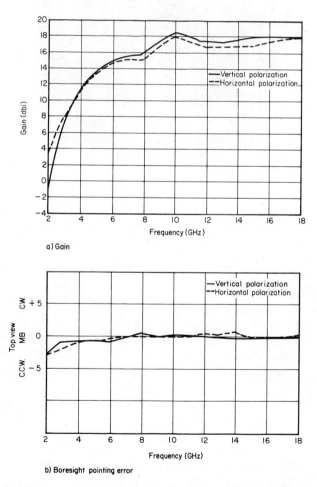

a) Gain

b) Boresight pointing error

Figure 3-57. Rotating reflector antenna (*a*) gain and (*b*) boresight pointing error.

Figure 3-58. Two axis shaped reflector scanning antenna.

fiberglass design with baffles that develop the required pencil beam at gain levels of approximately 35 dB. Dual servo systems provide the two axes of rotation.

3.6 SUMMARY

Chapter 3 presented the most popular types of antennas in current use for passive direction finding. It is not always clear why a certain design or system has been chosen, so emphasis has been placed on the key features of the designs. The actual antenna choice will generally be dictated by the system, the required accuracies, and the cost.

A DF system can almost be made as accurate as desired if sufficient antenna aperture is available. The driving factors, however, are the available geography for the DF system, the host vehicle requirements, and the mission needs. For simple determination of locations, line of bearing accuracies of 8–10 degrees peak are usually good. For targeting or aiming, accuracies of better than 2 degrees are required. For high-gain electronic countermeasures (ECM) or jamming, fractional accuracies are needed. In these cases the accuracy must be developed by signal processing both by analytical means and by digital or processing means. We shall study both methods in the chapters that follow.

REFERENCES

1. "IEEE Standard Test Procedures for Antennas," Antenna Standards Committee, ANSI Institute of Electrical and Electronics Engineers: New York, Wiley, 1979.
2. Turner, E., "Spiral Slot Antenna," U.S. Patent 2,863,145, Dec. 1958.
3. Booker, H. G., "Slot Aerials and Their Relation to Complementary Wire Aerials," *Journal Institute of Electrical Engineering*, Pt. IIIA, 1946, pp. 620 ff.
4. Kaiser, J. A., "The Archemedian Two-Wire Spiral Antenna," *IRE Transaction AP-8*, 1960, p. 313.
5. Rumsey, V. H., *Frequency Independent Antennas*, New York: Academic Press, 1966.
6. Dyson, J., "The Unidirectional Equiangular Spiral Antenna," *IRE Transaction AP-7*, Oct. 1959.
7. DuHammel, R. H., and D. E. Isbell, "Broadband Logarithmically Periodic Antenna Structures," *IRE Intern. Conv. Record*, 1957, pp. 119 ff.
8. Marchand, N., "Transmission-Line Conversion," *Electronics*, Dec. 1944, pp. 142 ff.
9. McLaughlin, J., D. Dunn, and R. Grow, "A Wide-Band Balum," *IRE Transactions on Microwave Theory and Techniques*, July 1958, pp. 314 ff.
10. Stutzman, W., and G. A. Thiele, *Antenna Theory and Design*, New York: Wiley, 1981, p. 283.
11. "Antennas, Masts, Mounts and Adapters," *AEL Catalog* 7827.5M, American Electronic Laboratories, Lansdale, PA, 1982, p. 46.
12. Hahn, G., and R. Honda, "Conical Spiral Arrays for Passive Direction Finding," *USAF Antenna Symposium Proceedings*, Allerton Park, Monticello, IL, Oct. 1968.
13. Kaiser, J., H. Smith, W. Pepper, and J. Little, "A Passive Automatic Direction Finder," *Proc. IRE 9th ECCANE*, Baltimore, MD, Oct. 1962.
14. Dyson, J., "The Equiangular Spiral Antenna," University of Illinois Antenna Laboratory, Report No. 21, Apr. 1955.
15. Chadwick, G., "Multiple Arm Spiral and Its Derivatives for DF and Homing," Presented at Los Angeles IRR Group on Antennas and Propagation, Jan. 15, 1970.
16. Lantz, P., "A Two-Channel Monopulse Reflector Antenna System with a Multimode Logarithmic Spiral Feed," Presented at the 16th Annual USAF Symposium on Antenna R&D, Radiation Systems, Inc., Report, NASA Contract NAS5-9788, Oct. 1966.
17. Chadwick, G., and J. Shelton, "Two Channel Monopulse Techniques—Theory and Practice," International Convention on Military Electronics, Sept. 1965.
18. Lipsky, E., "The Antector, a Millimeter Wave Spiral Detecting Antenna," Presented at the 3rd Annual Benjamin Franklin Symposium, IEEE MTT&AP Societies, Philadelphia, PA, April 1983.
19. Skolnik, M., Ed., *Radar Handbook*, New York: McGraw-Hill, Chap. 8.
20. Hopfer, S., "The Design of Ridged Waveguides," *IRE Transactions-MTT*, Oct. 1955, pp. 20 ff.
21. Jasik, H., Ed., *Antenna Engineering Handbook*, New York: McGraw-Hill, 1961, pp. 10–13.
22. Howard, D., "Single Aperture Monopulse Radar Multi-Mode Antenna Feed and Homing Device," *IEEE International Convention on Military Electronics*, Washington, D.C., 1964.
23. Honey, R., and E. M. T. Jones, "A Versatile Multiport Biconical Antenna," *Proceedings of IRE*, Vol. 45, No. 10, Oct. 1957.
24. Chubb, E., J. Grindon, and D. Venters, "Omnidirectional Instantaneous Direction Finding System," *IEEE Transactions on Aerospace and Electronic Systems*, Vol. AES-3, No. 2, March 1967.
25. Radio Research Laboratory, Harvard University Staff, *Very High Frequency Techniques*, New York: McGraw-Hill, 1947, Vol. I, p. 164.

26. Bohlman, W., "Broadband Rotatable Direction Finding," *Antenna AP-S International Symposium Digest*, 1977, p. 104.

27. Luh, H., and A. Tsao, "Null Shift and Beam Squint in Circularly Polarized Offset-Reflector Antenna," *IEEE Transaction AP-34*, Jan. 1986, p. 97.

28. Chu and Turrin, "Depolarization Properties of Offset Reflector Antennas," *IEEE Transaction AP-21*, May 1973, p. 339.

Note: The above list of references in antenna technology constitute only a small part of the available antenna literature. The author wishes to apologize to those not listed; the reader will find additional source material in the listing of references by the above authors.

Chapter Four

DF Receiver
Configurations

In this chapter, passive DF systems used for radar warning and ELINT purposes will be described in terms of the DF concepts and antenna technologies presented thus far. System configurations typical of present-day implementations will be detailed in various generic block diagrams. In each case, the reader should identify the type of system and, if monopulse, determine the class and method of satisfying the criteria for angle extraction. This recognition process will develop an appreciation of the potential a given system has to attain specific goals. Some systems are designed for warning, some for data gathering, and some for high accuracy; in each case, a specific design offers advantages and disadvantages. A wide instantaneous bandwidth receiver may be readily jammed; a narrow bandwidth system will lose data but may be more accurate measuring the data it does process. It is hoped that by following these rules, the reader will develop an insight into system "entropy," or overall capability to achieve a desired objective.

4.1 DF RADAR WARNING RECEIVERS

In the microwave region, angle-of-arrival data is used primarily as a means of determining the line-of-bearing of a signal or the position of an emitter with respect to either the DF host vehicle or the overall situational geography. The first case is usually for threat warning, where the identification, direction, and rate of closure are important factors for self-protection. The second case is for electronic order of battle (EOB) for the purposes of calibrating the signal environment, to determine the circle of error probability (CEP) or location, to recognize changes in the presence of known static situation, or the initiation of a new threat condition. We have studied the rotating DF system utilizing skirt-inhibition and its evolution to monopulse, which, because of its capability for high accuracy, addressed the EOB problem.

This has led to the study of monopulse theory as utilized in radar. To show the application of the theory to passive DF systems, it will be useful to study one of the most popular applications, that of the radar warning receiver (RWR).

Threat warning applications are characterized by the need for wide open frequency and angular coverage, rapid identification for determination of threat status, and DF information for self-protection. Figure 4-1 is a diagram of what we shall define as a generic radar threat warning receiver (1–4). It is a four-antenna system utilizing wide beamwidth, wide bandwidth spiral antennas to provide high probability of intercept over a typical 2–18 GHz frequency band.

Four antennas geographically dispersed about the host vehicle provide four angular quadrants of coverage of 360 degrees of azimuth. The type of antenna used for this application must have essentially no backlobes and must exhibit antenna beamwidths of approximately 90 degrees to achieve 360 degree omniazimuthal coverage. Planar cavity-backed spiral antennas, as discussed previously, closely exhibit these properties and for this reason are most commonly used in RWR applications. The idealized patterns of four such spirals is shown in Figure 4-2. Although these patterns vary in 3 dB beamwidth, from 70 degrees at 18 GHz to approximately 100 degrees at 2 GHz, full 360 degree coverage is provided unambiguously, with sufficient articulation to make measurements. The angle of arrival ϕ of an input signal measured from boresight by the receiver corresponds exactly to the case of an amplitude monopulse system as defined previously for the equivalent radar case. Since the beamwidths are wide as compared to radar, the accuracies and resolution are correspondingly poorer (typical values lie between 5 and 10 degrees). The formation of the amplitude monopulse ratio requires determination of the strongest and next adjacent strongest signals, shown in Figure 4-2 as V_A and V_B, a process easily implemented in fast parallel logic. Since the antenna output is RF signal energy, the receiver must be in close proximity to reduce losses, especially at the high end of the 2–18 GHz band. Each antenna is therefore connected to a receiver, or a dual receiver can service two antennas if the RF cable runs are short, as is the case in wing-tip or helicopter installations. Detection of the signal takes place for the frequency range in one band, or the detector may be preceded by a band separation multiplexer, which divides the band into sectors. Figure 4-3 shows these various possibilities. In Figure 4-3a, the entire 2–18 GHz band is detected and logarithmic amplified for each antenna. A diode limiter precedes the detector (shown for all cases) to prevent burnout at high signal illumination levels. This configuration gives the maximum intercept probability but suffers most from the possibility of pulse overlap or continuous wave (CW) problems in a dense signal environment, since no reduction of data by frequency band selection is possible. In Figure 4-3b, each of the four antennas is connected to a quadraplex filter, which divides the original 2–18 GHz band into sub-bands, each of which is detected and logarithmically amplified. This configuration provides the maximum probability of intercept and the feature of frequency band information as a result of the multiplexing. It is, unfortunately, the most expensive and complex, with 16 video signals to be handled by the bearing computer. The data rate is high but can be reduced by sampling the 16 outputs on a band-by-band basis. This switching can also be used to sample or "chop" CW

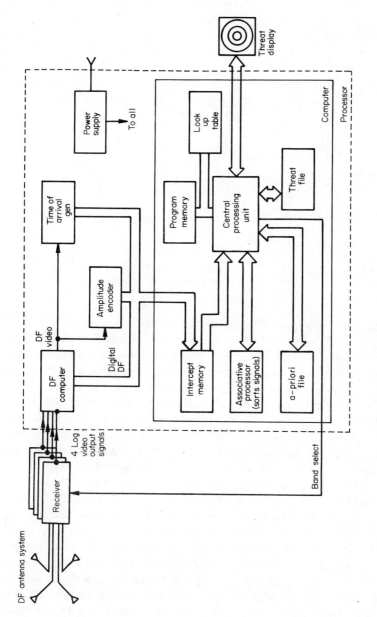

Figure 4-1. Generic warning receiver block diagram.

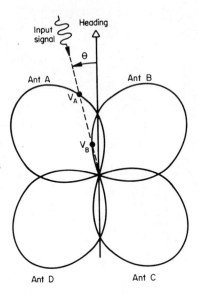

Figure 4-2. Idealized patterns for four geographically dispersed wide beam antennas.

signals to permit processing CW signals as if they were pulse types. The configuration in Figure 4-3c is an intermediate design utilizing quadraplexing of the frequency ranges in conjunction with switching to pick a band at a time. Some bands will be of greater priority than others, thus the switching need not always be sequential. Preamplification (which is a commonly used misnomer since it is actually postdetection linear video amplification) follows the detector to reduce distribution losses and improve sensitivity. This permits one or more bands to be paralleled in the switching process, which can be done when the environment is light, to improve intercept probability. True RF predetection amplification may be used; however, this is rarely done since the receivers must be well matched and may be located where different temperature extremes are encountered, making gain matching difficult.

Logarithmic video signals from the receivers are next fed to the DF computer (Fig. 4-1) where they are appropriately grouped by band and by antenna. A determination of the strongest and next adjacent strongest signal is made for antenna in each band. This establishes the two logarithmic signals that are to be used to form the monopulse DF ratio. The rule of adjacency eliminates backlobes and identifies an undesired time-coincident situation, such as two strong signals on opposite lobes, that may require data rate adjustment. It is assumed here that two acceptable strong and next strongest signals are available above a required signal-to-noise ratio, determined by a threshold setting. If only one acceptable signal is received just above the threshold in one channel antenna lobe, it may be ignored or assigned to boresight of that lobe.

The DF ratio data and the strongest signal is fed to the time-of-arrival generator and amplitude encoder to form a set of data for each pulse consisting of DF, amplitude, time of arrival, and pulse width. As each pulse is received it is fed to the intercept memory. This process is effectively similar to taking a snapshot in time of the signal environment with the exposure time analogous to the time (capacity) to fill the memory. When the memory is full, the encoded signals are either ignored or fed to another parallel memory until the first time snapshot is processed. This processing is accomplished in non-real time to deinterleave the pulse by separating them into pulse trains, after which they are compared to an a priori memory for identification. A look-up table provides DF correction as a function of band or azimuth if calibration data is available.

The result of the processes described above is a prioritized threat file that is displayed to the operator on a threat display. This is usually a polar CRT plot of signal type versus angle of arrival at range of imminency or danger, based upon signal strength, lack of scan, constant illumination, or identification with a priori data. After a cycle of processing, the intercept memory is cleared, the process

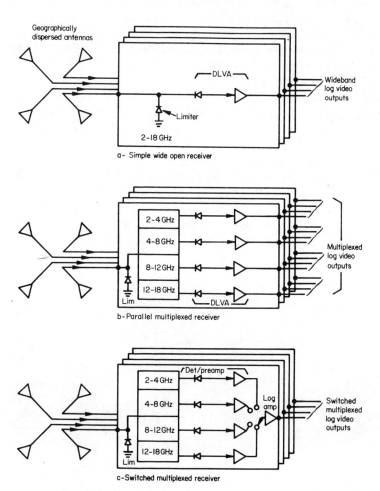

Figure 4-3. Various warning receiver front ends. (*a*) Simple wide open receiver. (*b*) Parallel multiplexed receiver. (*c*) Switched multiplexed receiver.

repeats itself, and the display is refreshed. The simpler generic radar warning receiver described above is defined as DF driven; that is, decisions are made on the basis of DF and pulse-time information. Frequency data, if available at all, is only on a band-by-band basis. Since arriving signals vary in strength, the monopulse signal-to-noise ratio may vary in amplitude. This may degrade the quality of the DF measurement, and care must be taken to select an operating threshold for the system that prevents a single signal from falsely appearing at several DF angles. If a signal becomes strong enough to overload the receiver, it can cause the DF measurement to vary again with the same result. For this reason, high DF accuracy and large dynamic ranges are necessary. This has been accomplished in recent years by improved detectors and video logarithmic amplifiers.

Figure 4-4 is a photograph of the APR-39A (XE-1) radar warning receiver, manufactured by the Dalmo Victor Division of the Singer Corporation, illustrating the mechanical configuration of a helicopter-mounted high-performance receiver utilizing special antennas and receivers, manufactured by American Electronic Laboratories, to cover frequency ranges extending from the H through M bands (5). Figure 4-5 is a photograph of another high-technology radar warning receiver, the General Instrument model AN/ALR-80, which also illustrates advanced RWR technology (6). In the AN/ALR-80 four spiral antennas that are geographically dispersed about an aircraft feed four receivers (shown in the center), each of which, in parallel, detects and amplifies the DF signals feeding them to the digital signal processor (far right). Here a complex computer capable of 1.2×10^6 operations per second determines the monopulse DF and associates signals to permit their identification against a directory of over 1000 known signals stored in an electronic erasable programmable memory (EEPROM). This data bank can be preflight updated in 90 seconds at a flight-line to accept different environments. Fifteen emitters of the highest priority are displayed on the digital display indicator (upper left) as alphanumeric symbols, designating the class of signal (friend, foe, or unknown) and the range

Figure 4-4. The APR-39A (XE-1) radar warning receiver. Courtesy of Dalmo Victor, Division of Singer Corp.

Figure 4-5. The AN/ALR-80 radar warning receiver. Courtesy of General Instrument Corp., Government Systems Division.

and DF bearing of each. The information is presented by placing the intercept symbol at a distance from the origin as a function of threat priority at the angular position at which it was received. A CD band blade antenna and power supply complete the design. The AN/ALR-80 covers 360 degrees of azimuth and measures DF bearing, either true or relative, to better than 15 degrees accuracy. It has a shadow time of 2 microseconds and is capable of self-test built-in-test equipment (BITE) for preflight checkout. This unit covers the entire frequency range from E to J bands in four bands.

4.2 PHASE AND SUM AND DIFFERENCE MONOPULSE DF RECEIVERS

Phase monopulse receivers make use of the phase difference between two signals to provide an angular output that can be either the time-difference-of-arrival in the case of the simple interferometer or the phase change of one antenna beam with respect to another in the general case. Since amplitude monopulse systems can be converted to phase type, it is often convenient to substitute one for the other when there are processing advantages. Since phase matching is more difficult to achieve than amplitude balance, most radar designs make use of sum-and-difference techniques, which in part shares the possibility for error by utilizing phase for ambiguity

resolution rather than absolute angular measurement. This often makes phase monopulse measurement popular for multiple baseline interferometer systems as a means
to resolve nulls. It is useful to consider some simple and unusual phase and combinational phase/amplitude monopulse applications to illustrate the technique.

Figure 4-6 shows the simple interferometer described in Chapter 2 connected to
the class III processor. Repeating Equation (2-10),

$$\psi = \frac{-2\pi D}{\lambda} \sin \phi \qquad (2\text{-}10)$$

and Equations (2-11) and (2-12),

$$A = M \sin(\omega t) \qquad (2\text{-}11)$$

$$B = M \sin\left(\omega t - \frac{2\pi D}{\lambda} \sin \phi \right) \qquad (2\text{-}12)$$

(the negative sign indicating a time lag). This shows that the frequency ($1/\lambda$) must
be known to determine the angle of arrival. The class III processor measures the
phase angle difference between A and B, yielding ϕ. All amplitude variations are
removed by the limiter in the process. Since it is not possible to know whether A
leads or lags B, there is a 180 degree ambiguity in the measurement. Practical
systems use a third channel to resolve this problem. This restriction generally limits
the class III processor to narrow field-of-view systems, chiefly for scanning radar
applications.

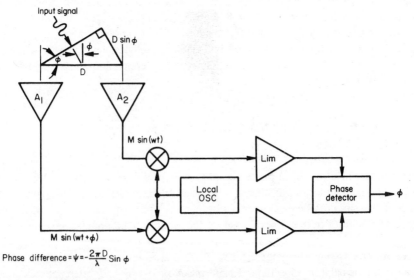

Figure 4-6. Simple interferometer connected to a class III processor.

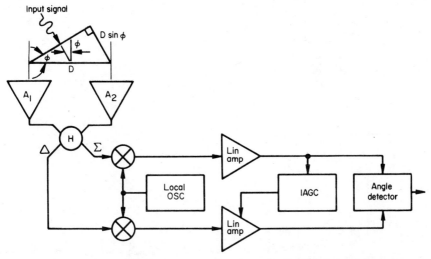

Figure 4-7. An RF sum (Σ) and difference (Δ) interferometer used with a class I processor.

Figure 4-7 is the RF sum (Σ) and difference (Δ) case of the interferometer, which we know as the phase monopulse detector. The use of a hybrid at RF yields two signals that contain the angle-of-arrival information in *both* the amplitude and phase of the two detected signals. Ambiguities in the phase measurements can now be resolved over 360 degrees by the additional data that can give quadrant (amplitude) as well as phase difference data. This is a system that can be practically used over a wide bandwidth, limited however by the presence of phase ambiguities for signals with a wavelength of less than $D = \lambda/2$. Interferometer methods will be examined in Chapter 6, which discusses interferometer techniques to resolve the above ambiguities.

Phase monopulse DF systems are often used in conjunction with amplitude comparison monopulse techniques (7). An excellent example of this is the use of a four-arm cavity-backed planar spiral antenna to provide both the azimuth (ϕ) (*E* plane) and elevation (θ) (*H* plane) angles of an intercept on a pulse-by-pulse basis. This technique makes use of the multiple mode properties of the spiral antenna to operate simultaneously in several radiation modes.

Figure 4-8 is a diagram of the four-arm spiral spatial monopulse direction finder antenna and feed configuration that was described in Chapter 3. The outputs of the combination are sum (Σ) and difference (Δ) signals that contain all the information required to develop both the azimuth angle ϕ and elevation angle θ over a full hemisphere. The amplitude ratio of the (Σ) and (Δ) patterns yield θ; the differential phase between (Σ) and (Δ) contain ϕ. Although the ratio of the amplitudes of the sum and difference patterns will give the angle θ between the two, it will not yield the *H*-plane elevation angle since its magnitude is measured off boresight in *any* direction and the patterns are symmetrical on either side. Knowledge of ϕ, however, resolves this difficulty, permitting the correct elevation angle to be determined. The

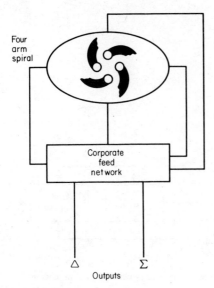

Figure 4-8. Four-arm spiral spatial DF system with $\Sigma - \Delta$ output.

problem is to determine both the phase and amplitude differences between the sum and difference signals accurately to extract the required DF information.

The processor required to develop the azimuth and elevation angles, shown in Figure 4-9, uses a two-channel superheterodyne receiver as opposed to the standard three-channel radar processor, which would compare the sum to each of two difference signals for azimuth and elevation. A linear amplifier brings the sum and difference signals to a level suitable for distribution to the class II logarithmic IF amplitude

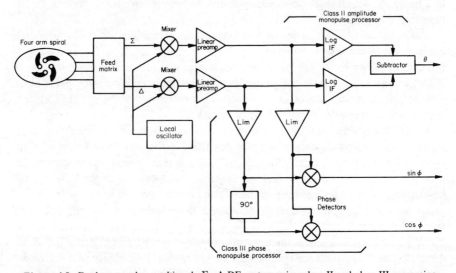

Figure 4-9. Dual monopulse, multimode $\Sigma - \Delta$ DF system using class II and class III processing.

comparator and to the two class III phase comparators. Dual limiters and a 90 degree phase shifter in one channel permits both the sine ϕ and cosine ϕ to be determined to permit monopulse measurement of tan ϕ and cot ϕ simultaneously (tan $\phi = 0$ requiring determination of cot ϕ and vice versa). This gives the azimuth angle of arrival ϕ. The subtractor yields θ, the angle of elevation.

This example is only one of many combinational systems when the difficulty of balancing phase or amplitude leads to the use of one technique to aid the other. In general, for narrow bandwidth superheterodyne systems, a pure amplitude monopulse comparison is not as accurate as a pure phase monopulse system, with the combinational system accuracies lying in between. For wide instantaneous bandwidth systems amplitude comparison monopulse systems are generally more accurate than phase types.

4.3 SUBCOMMUTATION METHODS

4.3.1 Pseudomonopulse Receiver

During the development of monopulse as a parallel-lobe radar receiver technique, it became evident that a switched-lobe method might have some advantages in simplifying the receiver (8). If we consider Figure 4-10, it is apparent that a single-channel receiver could be switched from antenna lobe A to antenna lobe B on a pulse-by-pulse basis. The input switch SW1 can be synchronized with the output switch SW2 permitting the two logarithmic voltages V_A and V_B to be stored in integrating capacitors C_A and C_B. It is then possible to take the required ratio, logarithmic V_A/V_B, at the output of the amplifier connected to the integrating capacitors some time after both capacitors have been charged. This type of circuitry is known as a sample-and-hold or boxcar detector. Detection is performed on a pulse-by-pulse basis to permit the capacitors to be discharged periodically, readying them for the next pulse.

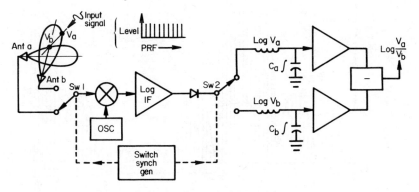

Figure 4-10. Switched-lobe pseudomonopulse system.

The switched-lobe technique works well in radar receivers since the scanning process gangs the transmit antenna with the receiver antenna thus removing the effects of the radar scan. Since the transmitted signal is a constant amplitude (constant illuminating) train of pulse signals, the amplitude difference between the received lobes is due to the ratio of amplitude of the signal received in each beam, with equal signals denoting boresight. The technique is also used for the detection of CW signals by developing the switching or pulse train by a clock, which is later used to demodulate the resultant "video" in AC coupled amplifiers. Subcommutation is often referred to as lobe-on-receive-only (LORO) or pseudomonopulse and, despite the pulse-to-pulse delay, is considered a form of monopulse since the major amplitude variation is due to the angular information derived by the received beams. In a reasonably busy environment, signals are mixed: Long-range surveillance radars may scan with periodicities of many seconds; short-range radars may have kilohertz scan rates; there may be constant illumination monopulse and high PRF doppler and CW radars. Subcommutation would appear to have a problem detecting scanning type radars, however; fortunately, a solution is implicit in their fundamental radar design. All radars have to satisfy a basic design equation, which requires that a significant number of pulses be returned from the target to permit detection at the greatest desired radar range, limited by the duty cycle of the radar transmitter. This dictates a PRF that is many times the inverse of the antenna scan rate, thus placing many pulses under the scanning antenna beamwidth. For the passive subcommutation receiver, it is possible to assume a *maximum* rate of change of pulse amplitude as a result of these factors and to derive an "omni" or reference channel artificially to remove the target scan. This is best illustrated in Figure 4-11, which is an exaggerated drawing of a typical scan radar. The figure depicts the pulse-to-pulse variation bound by assuming a number of pulses that the scanning radar needs for range determination. This number is a fraction of 1 dB per pulse in L band ranging to

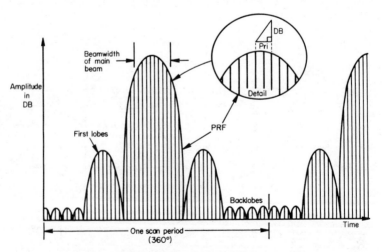

Figure 4-11. Received intercept showing PRF content in scan.

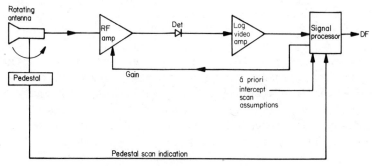

a) Simple rotating antenna pseudo monopulse DF system

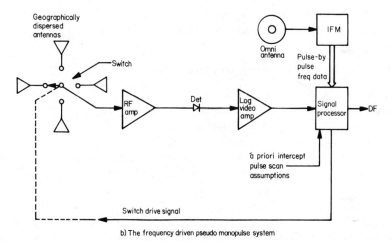

b) The frequency driven pseudo monopulse system

Figure 4-12. Two forms of a single-channel subcommutated "pseudomonopulse" system. (*a*) Simple rotating antenna pseudomonopulse DF system. (*b*) Frequency-driven pseudomonopulse system.

several decibels per pulse at X band and higher frequencies. A rule of thumb often used assumes a minimum of 30 pulses per 10 dB antenna beamwidth assuring a maximum of about 1.5 dB variation from pulse-to-pulse for the worst case.

The simplest pseudomonopulse receiver is merely a single channel composed of a rotating DF antenna, detector, and video amplifier and a processor as shown in Figure 4-12a. By programming the gain of the system and comparing the known scan-to-scan information developed by the rotating DF antenna, it is possible to develop a series of pulses-per-scan that can be time associated into pulse trains by assuming relatively constant pulse repetition intervals (PRI). The amplitudes of each train can then be compared to determine DF and corrected for the expected target scan as described above. The scan-to-scan gain programming is necessary to assure that the signals are within the dynamic range of the receiver. It is also possible to eliminate backlobe response with this method by only using the strongest lobe of a scan that is known to be unsaturated. This can be determined by examining scan-to-scan data with a computer. Although this system can and has been made to work, it is not useful in dense environments since it takes excessive time.

A more practical type of subcommutated pseudomonopulse receiver substitutes multiple antennas switched on a pulse-by-pulse basis as shown in Figure 4-12*b* and as described for radar receivers in Ref. 9. The input signal is received by an omniazimuthal antenna feeding an instantaneous frequency measurement (IFM) receiver that provides frequency data on each pulse. The signal processor associates the pulses into pulse trains by frequency and forms the DF ratio of the signal amplitude received on one antenna at time t_1 with the same frequency signal received on the next adjacent antenna at t_2, where $t_2 - t_1$ is the derived pulse interval. Switching is done on a pulse-by-pulse basis. The pseudomonopulse ratio thus obtained is corrected by a priori knowledge of the intercept expected radar scan. The DF output of this system has, in effect, been normalized removing the intercept scan ratio to the extent of the accuracy implicit in the scan prediction.

4.3.2 True Monopulse Subcommutated Receiver

Figure 4-13 is a diagram of a true monopulse frequency-driven subcommutated receiver. An omnidirectional antenna feeds an IFM receiver that makes pulse-to-pulse measurements of the signal environment, feeding this information, grouped with analog signal data, to a processor where digital frequency, pulse amplitude, time of arrival, and pulse-width data fields are formed. An eight-element antenna array (shown here as two 45 degree offset-switched four element pairs of spirals) is switched as a *sequenced pair* providing two adjacent signals that develop a series of logarithmic ratio signals that are temporarily stored as DF angle-of-arrival possibilities (10). These analog logarithmic ratios are also provided to the processor for time-of-arrival encoding. Once a set of pulses is frequency associated into a train by data from the IFM subsection, it is possible, by time-of-arrival comparison, to make the DF measurement, the frequency word acting as a filter to associate pulses into train using the time-of-arrival word to deinterleave the DF pairs from the environment. The DF is a true monopulse measurement, subcommutation being used as a means to reduce the number of channels. High intercept probability is provided by omniazimuthal IFM channel; however, DF measurement is slow, often requiring a decision as to whether or not to make it based on frequency and/or other omni-measured characteristics.

In the system shown above, low-gain wideband antennas, such as planar spirals, are paired to obtain the true monopulse ratio eliminating the correcting assumption and requiring that each pulse be received in *both* antennas of the pair. This reduces the system sensitivity compared to the single-channel pseudomonopulse technique depending upon processing assumptions. Broadband RF switching is used, necessitating the use of matched switches. Two RF amplifiers are often added, to improve sensitivity of the DF channel making the switched-pair system sensitivity approximately equal to that of the omniazimuthal IFM receiver channel. In another configuration a number of narrow-beam octave horn antennas can be used to obtain higher sensitivity.

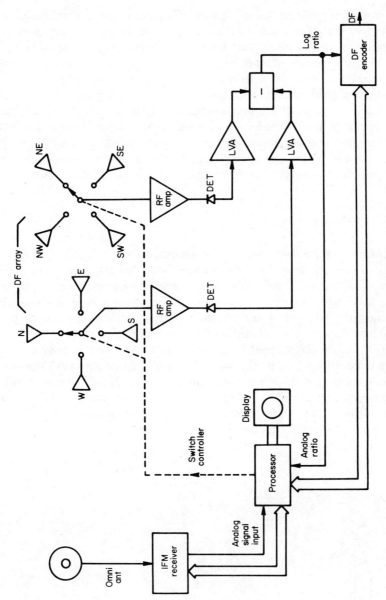

Figure 4-13. Subcommutated frequency-driven receiver.

The improvement gained is due to the relative antenna gain of a horn compared to a spiral, permitting the video switching concept to be used at the expense of instantaneous frequency bandwidth, since the horns do not provide as wide an RF bandwidth. The sequenced offset-pair system shown is typical of a relatively complex subcommutated system, the chief virtue of the paired technique being the true amplitude monopulse DF normalization. It may also suffer from the physically different field of views of the omniazimuthal and DF antennas since they must usually be placed at different locations. Some systems make use of a monopulse pair formed by the omni antenna and one switched antenna of a DF array. The problem with this approach, however, is the different gain and field-of-view between the antennas due to construction, polarization, and physical displacement.

4.4 SUPERCOMMUTATION

The problems in subcommutation can be overcome, in part, by the technique of sampling or supercommutation. This is defined as the process of switching to each of n antennas during the duration of a given pulsewidth for the purposes of obtaining the monopulse ratio on a single pulse (11). This technique is most nearly equal to monopulse in the formation of the ratios and has the incidental advantage of chopping CW and high-duty cycle signals by the high-speed switching action.

Figure 4-14 is a diagrammatic representation of the technique. Four antennas, having a finite RF bandwidth, are sampled by a high speed RF switch capable of being activated at a rate that allows time multiplexing of each of their outputs within a single received pulse width. The resultant signal is amplified by a common RF amplifier, detected, and logarithmically amplified. Deinterleaving or separation of the original four signals is accomplished by a decommutator, synchronized with

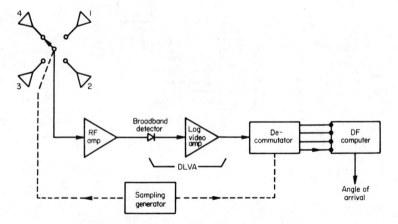

Figure 4-14. Supercommutated DF receiver.

the RF switch, permitting recovery of the four original video signals, which are then formed into the appropriate amplitude monopulse ratio of strongest to next adjacent strongest signal. The ratio thus formed is the familiar amplitude monopulse data, which can then be processed as before.

Ordinarily, supercommutation must be done at a rate consistent with the Nyquist criteria, which states that if the occupancy of the pulse signal band is limited to B_I, the rate must be $2B_I$. This would typically require a 20 MHz rate for a 10 MHz wide occupancy pulse signal. The need to interleave four signals, however, necessitates the use of a rate consistent with the rise and fall times of the interleaving switching, which is at least four times faster (for a four-antenna system). Typical bandwidths of 80–200 MHz are therefore used for video and logarithmic amplifiers in systems of this type.

Supercommutation, as a technique, allows multiplexing of RF signals by time division. The isolation between any two adjacent channels determines the maximum difference of RF amplitude that can be accommodated. If, for example, 25 dB of isolation is obtained, then the range of signal in one beam of an antenna to that of the next adjacent one could use this value, allowing a high articulation or rate of pattern change-per-degree of arrival angle, which is desirable for maximum accuracy. The limiting factor in obtaining high isolation is the time it takes to go from one arm of the switch to the next. The limit to the number of channels to be supercommutated becomes the dwell time and change-of-state time for the switching. For this reason, practical systems have been confined to four or five channels. With the development of faster switches, the technique could be expanded to multihorn arrays for example. The combination of the multiple input channels, incidently, eliminates the problem of RF combining that would occur if adjacent antennas were RF summed.

In the supercommutated system, as shown in Figure 4-14, the four video signals are recovered completely before the processing subsystem is used to determine pulse parameters, time of arrival, and monopulse direction. This requires that an analog-to-digital conversion take place late in the system to permit digital processing for signal recognition. In the generic warning receiver, a signal processing computer would form a digital word or field for each signal. For example, a typical intercept might have an 8-bit word for angle of arrival and signal amplitude, a 16-bit word for arrival time, another 8-bit word for antenna beam and pulse probability characteristics, and, in some receivers, a 16-bit word for frequency. These conversions would be made and the resultant digital words loaded into storage for processing, usually dynamic random access memory (DRAM) is used for this purpose, making the storage a time-consuming process.

Supercommutation develops much information in a form that can be processed rapidly by virtue of comparisons that can be made in near real time. Figure 4-15 shows an adaptive supercommutated DF system that performs signal preprocessing in advance of loading of the memory of the computer for the purposes of saving analog-to-digital conversion and storage loading time. In the hypothetical system shown, all data developed by the supercommutation process is reconstituted into "parallel compare" channels and fed to a preprocessor comparator where DF, signal

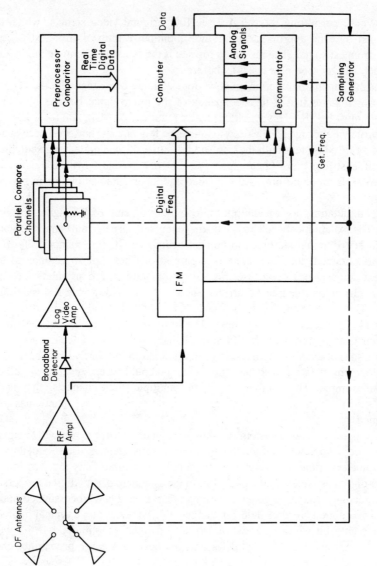

Figure 4-15. Adaptive supercommutated preprocessor DF system.

amplitude, and other information can be extracted during the duration of the incoming pulse. This can be done by structuring the architecture of the preprocessor specifically for this purpose. Since it is possible to make comparisons faster with high-speed gates than it is to perform an analog-to-digital conversion, load a random access memory, and then do the processing; time is saved. If DF driven, monopulse DF decisions can be used to stop the supercommutation switch at the antenna receiving a desired signal, feed the RF from this antenna to the instantaneous frequency receiver (IFM), and then develop a digital frequency word. Since IFM frequency measurement is usually the slowest part of the process, only pulses that meet the direction-of-interest criteria will be measured, reducing the shadow time of the receiver substantially. The decision-making process in this system, however, is being performed on a single or monopulse requiring very large signal-to-noise ratios to reduce false alarms. This requirement is fortunately in concert with the ability to make the crystal receiver highly sensitive or noise-limited by use of a single linear RF amplifier. The single-channel time multiplexing eliminates channel-to-channel balancing that would be required in a multichannel parallel system; however, the switch balance, isolation characteristics, and dynamic range capabilities must be carefully controlled.

The preprocessor supercommutated system is a compromise between sensitivity and total intercept probability since it sacrifices the higher sensitivities of the narrower video bandwidth amplified crystal video receiver for the feature of direct comparisons (true monopulse) and the ability to measure frequency selectively. It exhibits only moderate operating sensitivity, in the order of that of a standard radar warning receiver. The availability of the frequency word to act as a sorting tool, however, is a desirable feature. The system is also limited by finite switching capabilities, placing a limitation on the minimum pulse width that can be used. As a practical matter, the adaptive supercommutated preselector receiver is a good example of the trade-offs that can be made between relatively low-cost digital componentry and expensive RF circuitry.

The advantages of supercommutation, which is actually time-division multiplexing (TDM), is the relative simplicity and balance improvement obtained by the use of a single amplification and detection channel for all four directional signals. The occupancy bandwidth of the switching processes, however, makes the system subject to problems of noise, which must be considered in any system design trade-off since the commutation process generates noise and harmonics at the rise-time rate of switching, which could generate harmonics that can appear in the RF passband of the receiver. The supercommutation technique does yield data on a single pulse basis, therefore placing it in the category of a true monopulse system.

4.5   MONOCHANNEL MONOPULSE

One of the major problems in developing monopulse receivers capable of covering an instantaneous 360 degree field-of-view is the need to replicate as many receive channels as there are antennas to form the monopulse ratio of strongest to next strongest signal. The problem is made more acute than that of channel cost by the need to match the amplitude and, in some cases, phase of the channels over a wide dynamic range. In the approaches presented to date, we have studied both subcommutation or pseudomonopulse and supercommutation or time-shared multiplexing as usable techniques; there is an analogous phase equivalent to supercommutation that merits our attention. This novel form of monopulse detection, developed by Watkins Johnson (12), utilizes a single shared channel and makes use of many of the principles of the time-multiplexed supercommutator without the disadvantages of expensive commutation and decommutation. It accomplishes this by phase-tagging each of the four (or more) amplitude DF signals such that a single super-heterodyne channel can process the resulting composite signal, thus eliminating balance problems as before.

To do this requires that each of the four signals to be downconverted be uniquely identified during the duration of the pulse width. In a supercommutated system this is done by the time position of a signal; that is, the RF switch that samples each channel is time synchronized with the video distribution switch to charge a capacitor to deinterleave that channel thus recovering the DF data. In the monochannel technique the phase state of the local oscillators that downconvert each of the channels differs from each other in a known way to identify or tag that channel. Knowledge of this coding is then used in the IF-to-video conversion to separate the signals in the demodulation process to recover the original channels.

Figure 4-16 is a simplified diagram of the technique. Each channel of a four-spiral antenna system is downconverted to a convenient IF by a double-balanced mixer fed by a local oscillator signal from each one of four biphase modulators M_1, M_2, M_3, and M_4. The biphase modulator introduces a pass (1) or no pass (0) state to the local oscillator signal essentially creating a phase-code modulated signal. The code is a maximum linear sequence that makes each local oscillator signal appear as a pseudorandom signal with an occupancy of the form of sin x/x. The code generator generates this sequence, which is then delayed or time-shifted by the delays t_1, t_2, and t_3, these delays providing the signal isolation necessary to tag each of the RF signals. The single channel, which now contains the four channels of DF data, is amplified and brought down to a convenient IF where the coded RF signal is then used to decode the phase state of the phase-modulated composite signal back into the four original DF signals, each of which can then be processed to extract DF data.

The above method is a true monopulse technique to the extent time supercommutation is, since the code rate must be selected to allow four states of phase to be obtained within a pulse width. Typical code rates are 200 MHz for a 200 nanosecond pulsewidth. At this value, there is approximately 15–18 dB isolation between channels, which is sufficient to permit the use of spiral antennas since approximately 12 dB of signal difference occurs between two spirals at boresight when adjacent channels are compared (see Chapter 9, squint accuracy discussion). The monochannel technique can also include a fifth channel, an omniazimuthal for use with an IFM, for example. This would require readjustment of the rates.

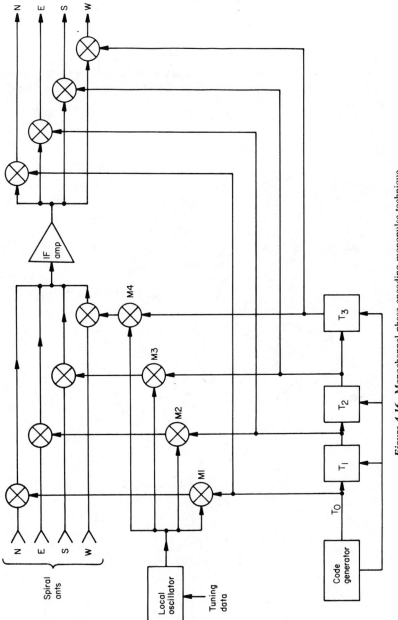

Figure 4-16. Monochannel phase encoding monopulse technique.

The significant advantage, outside of the obvious cost saving of the monochannel approach, is that it again permits combination of the four RF signals into a single channel without the problems of interferometer nulling or the generation of "grating lobes," which would occur if direct RF addition were attempted. The technique is most suited to relatively narrow bandwidth scanning superheterodyne receivers compared to supercommutation, which is best applied to wide RF bandwidth noise-limited crystal video receivers.

4.6   PARALLEL DF CHANNELIZATION

Parallel DF channelization DF receivers make use of array techniques or multiple antennas to provide wide-angle DF coverage. We will be covering arrays as a separate topic in Chapter 5; however, for the purposes of illustration of the technique consider Figure 4-17, showing the parallel arraying of individual antennas utilizing backlobe inhibition. The system shown is well suited for applications requiring

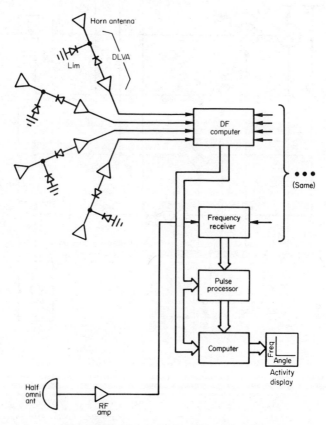

Figure 4-17. Parallel DF channelization receiver.

dispersed antennas, although single-point arrays are also in use. Each channel in the configuration shown makes use of a high-gain antenna limiter, detector, and video amplifier channel to provide coverage of a specific angular sector. Logarithmic amplifiers are used in each channel to permit direct comparison of all channels. Two half-omnidirectional antennas each covering approximately 180 degrees of view are used to provide backlobe inhibition of the horn array. These antennas conveniently provide an RF output to feed an instantaneous frequency receiver (p. 78).

The parallel DF channelized receiver offers many advantages, the chief being full omniazimuthal coverage at all times. The system can be either DF or frequency driven and can be readily coupled to an equivalent jamming system for both retro-directive and repeater jamming. The system also offers an immunity to jamming or overloading by strong CW emitters, since at the worst, only one or two sectors might be disabled. The parallel channelization concept is a powerful approach to jamming immunity. Although simple horn antennas are shown here, array techniques utilizing Butler matrixes, Rotman-Turner lens feeds, and ferrite phase shifting are in wide use and will be described in the next chapter. Since systems of this type are generally large, the output display is expanded to provide ELINT as well as threat information, giving this type of output an activity view of the electromagnetic environment around the host vehicle.

4.7 MULTIBEAM DF ARRAYS

Parallel multibeam array DF systems make use of an antenna system that provides a contiguous pattern coverage either for a sector or over a full 360 degrees. These types of systems are readily made conformal to the host vehicle as shown in Figure 4-18, which is one 90 degree sector of a 360 degree coverage system. In actuality, in a shipboard installation, two DF units would be used to cover 180 degrees on the port side and two more to cover 180 degrees of azimuth on the starboard side. In the system shown, the 16-beam multibeam array identifies the angle of arrival in just one 90 degree sector. The half-omniazimuthal antenna, which is similarly polarized and placed in close proximity to the DF array(s), feeds an IFM receiver to obtain frequency data. A digital tracking unit develops monopulse DF information, time-of-arrival, and frequency, all of which are grouped into threat words and fed to a computer for threat identification. This information is displayed to an operator who can take appropriate countermeasures, ranging from dispensing of chaff in the appropriate direction to application of ECM for larger systems. In this latter case, a series of travelling wave tube (TWT) amplifiers can be supplied the angle-of-arrival data directly and their outputs combined in a lens array to transmit a signal in the same direction as that of the input pulse. The two 90 degree sector 16-beam antenna systems combined with fore and aft beamformer antennas can be directly switched to a 35-multibeam array network fed by as many TWTs as there are beam ports. A modulated signal return would then be generated, amplified, and directed to retrodirectively jam the target, using appropriate modulation modes. The power

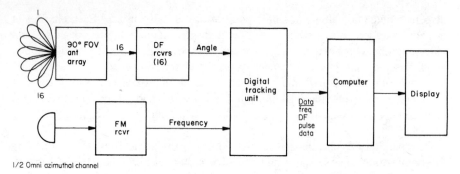

Figure 4-18. Typical multibeam conformal DF system.

of all of the TWTs would add up to place high effective radiated power (ERP) on the target. The advantage of the system is easily seen from the following:

$$P_j G_{Aj} = N^2 G_e P_A$$

where

$$P_j G_{Aj} \quad = \quad \text{the effective radiated power ERP}$$

$$N \quad = \quad \text{the number of array elements}$$

$$G_e \quad = \quad \text{the antenna element gain}$$

$$P_A \quad = \quad \text{the power output of a single amplifier}$$

The actual power on the target increases as the square of the number of array elements. This permits low-power TWTs (50 W level) in a practical 16-beam array, with typical 10 dB element gain, to transmit 100 kW power levels. There is also a redundancy feature since the failure of any one tube will not seriously degrade the ERP. This example explains the popularity of the Rotman phased array antenna ECM system, which is finding wide use. Passive detection systems also find ready application to targeting and aiming of ordinance where active radar is not to be used. The multibeam approach permits high precision to be obtained for this purpose as well.

4.8 SUMMARY

This chapter has presented the evolution of DF from a simplistic methodology to a more sophisticated one, utilizing the methods of monopulse radar and phased array technology. The important differences between the monopulse sensors and processors has been presented to show how interesting the differences are and how they really permit DF technology to be considered as art by itself. Examples of key systems have been used to present the more useful approaches, within the confines of previously released information. This practice will be followed throughout this book to familiarize the reader with the various forms of DF technology. It will become apparent that despite the many different paths passive microwave direction finding has taken over the years, actual systems really converge into fundamental types: rotary DF, commutated systems, interferometers, and parallel fixed antenna arrays.

REFERENCES

1. Belk, J., J. Rhodes, and M. Thornton, "Radar Warning Receiver Subsystems," *Microwave Journal*, Sept. 1984, p. 199.

2. "Egypt Deploys British ESM," Staff Report, *Electronic Warfare*, Defense Electronics, Vol. 10, No. 1, p. 79, Jan. 1978.

3. LaTourrette, P., "Multiplexers for ESM Receivers," Conference Proceedings, Military Microwaves, the Institution of Electrical Engineers and Institution of Electronic and Radio Engineers, Oct. 1978, pp. 9 ff.

4. "U.S. Steps Up Its Power Management Program," Staff Report, *Electronic Warfare*, Vol. 6, No. 3, May/June 1974.

5. "Digital Threat Warning System Completes Tests, Nears Army Contract," *Aviation Week and Space Technology*, May 6, 1985.

6. "The AN/ALR-80 RWR System," Data Sheet, General Instrument, Government Systems Division, Hicksville, NY.

7. Bullock, L., G. Oeh, and J. Sparagna, "An Analysis of Wide-Band Microwave Monopulse Direction Finding Techniques," *IEEE Transactions on Aerospace and Electronic Systems*, Vol. AES-7, No. 1, Jan. 1971, p. 189.

8. Page, R. M., "Accurate Angle Tracking by Radar," NRL Report RA3A222A, Plate 1, Dec. 28, 1944. Republished in *Radars Vol. I, Monopulse Radar*, Dedham, MA: Artech House, 1977.

9. Stone, T. W., "IF Commutation Increases Monopulse Radar Accuracy," *Microwave Systems News & Communications Technology*, Dec. 1985.

10. Lipsky, S. E., "Method and Apparatus for Electrically Scanning an Antenna Array in a Monopulse DF Radar System," U.S. Patent 4,313,117 granted Jan. 1982.

11. Lipsky, S. E., "Supercomponents Solve New DF Design Problems," *Microwaves*, Sept. 1975.

12. Klaus, D., and R. Hollis, "Monochannel Direction Finding Improves Monopulse Techniques," *Defense Electronics*, Vol. 14, No. 3, Mar. 1982.

Chapter Five ————————————————

DF Antenna Arrays

Array technology for passive direction finding has been developing as an outgrowth of two factors: first, the need to respond retrodirectively (transmit in the same direction as received) to a threat for jamming purposes; second, the need to measure a signal to a high degree of accuracy for targeting or aiming purposes. One of the main concepts of the field of electronic support measures (ESM) is the use of the electronic receiving systems as a means to detect signals for threat determination optimally and as a method of improving or multiplying the effectivity of jamming or electronic countermeasures (ECM). These functions are a subset of the concept of electronic warfare. With the advent of phased arrays, passive direction finding has again stolen a page out of the handbook of radar design.

Early radar systems first utilized scan then monopulse passive lobing in the receiver to illuminate and detect targets. As the number of targets (and radars) in a given environment increased, it became obvious that improved resolution in all areas would be needed. Doppler, moving-target indicators, and range-gating techniques were honed to a high degree of perfection to improve range resolution, and monopulse improvements normalized the radar signal returns, reducing false clutter and glint variations. Despite these changes, it became obvious in the early 1950s that fundamental changes would be needed in the antenna systems to improve angular resolution in both azimuth and elevation. Additionally, the requirement for improved security for countermeasures immunity and the ability to be able to adapt to changing conditions of the environment acted to advance the concept of a narrow beamwidth rapidly "steered" antenna. Arraying or essentially combining the effects of many elemental radiators to form one or more beams developed as a technique that could accurately direct or position a beam in free space, the arraying and combinational effects permitting development of antennas that could generate narrow beamwidths with high gain that could be switched or scanned. Since the maximum range of a radar is directly proportional to the effective radiated power (ERP), which in turn depends upon relative gain over an isotropic radiator, narrowing of the antenna beamwidths acted to improve range over the limits imposed by available transmitters and receivers.

A corollary advantage to the gain, obtained at the expense of reduced beamwidths in radar transmitters, was the distributed power concept inherent in any multielement radiating system. Simply stated, since there are many radiators, it is possible to connect each one to a separate source of RF power each of which, if made phase coherent, can effectively add its power in space, in contrast to the problem of generating a single source of high power and connecting it to a steered antenna system. The advantages of this distributed source technique become more important with solid-state oscillators. High power solid-state transmitters are still unavailable at high microwave frequencies, yet moderate power levels are obtainable from solid-state oscillators and amplifiers, which when properly phased and added, can create the desired high ERP necessary to achieve the required range improvements.

There are many phased array techniques in use at the present time that are of little interest in passive direction finding since they are highly dependent upon a priori knowledge of the transmitted signal. We will confine our consideration to types that require no such data. Phase arrays permit the accurate angular reception of signals by means of a high-gain narrow-angle beam scanned over the field of view (FOV) the array is designed to cover. A simple case is the planar array of multiple elements spaced in a straight line and fed in a phase, amplitude progression, or time sequence to effectively switch or scan over azimuth (ϕ) and elevation (θ) in a prescribed manner. By planar, it is understood that all radiating elements lie in one plane and that desired operation occurs only over a limited angle, typically ± 60 degrees from the perpendicular boresight or pointing angle of the plane of the array. If greater angular or hemispherical coverage is desired, a multidimensional array can be constructed by mounting assemblies of planar arrays about a host vehicle conformally or by constructing a four-faced pyramidal structure with an array mounted in each planar face. Each approach has its features: A conformal array encloses the host vehicle minimizing its shadowing or interfering effects; a stand-alone array minimizes problems of phase coherence and equalization of power feeds and centralizes the system for repair and maintenance. Another approach to hemispherical coverage is to eliminate the planar concept and mount the radiating elements cylindrically on the surface of a circular surface of sufficient diameter to permit appropriate phase and amplitude correction to be made to compensate for the nonplanar configuration.

In passive direction finding, use is often made of another implementation of the phased array, known as the adaptive array. Since it is desirable to receive a signal with the best signal-to-noise ratio, any information or decision that can steer or point the array beam in the optimum direction in anticipation of the signal will be of great help. The ability to reject undesired signals at different angular positions, with respect to the array, is also advantageous. The capability of programming an array offers potent features for future systems as computer decision making takes over the operation of DF systems.

5.1   PASSIVE DETECTION ARRAYS

It is useful to reduce the many types of phased arrays as applied to radar to the more specific requirements of passive direction finding. Our criterion here will be the application test; the array technology types that have proven useful in passive direction finding and have been generally reduced to practice will be considered since it is the purpose of this book to present a picture of applied technology. This limitation should not preclude future applications of new technologies such as state of the art microwave feed-line methods, RF signal generation, and RF switching; however, it is generally found that progress in electronics more often follows the route of evolution, not revolution (with some exceptions). It is reasonable to expect that future phased array methods for direction finding will gradually evolve from a logical progression of present technology.

There are two basic groups of phased arrays in current use: parallel-beam or high probability systems and switched-beam or directed systems. These are categorized in Figure 5-1. The parallel beam group is distinguished by the fact that all of the beams of the DF system exist at all instants of time; therefore, the antenna system provides an instantaneous angular coverage equal to its field-of-view, offering the advantage of high intercept probability. For circular arrays, this field of view is 360 degrees in the ideal case, providing omniazimuthal coverage. The directed-beam grouping provides a single narrow high-gain beam that can be steered or pointed in a given direction either as part of planned scanning process such as a raster scan, randomly as a function of time, or in accordance with a frequency program based upon knowledge of an expected return. In general, the steered-beam approach can be made to provide sufficient gain to look into the back lobes of emitters, which compensates, in part, for the reduced probability of detection due to the scanning process. The advantage of the electronically steered-beam is its inertialess scan, eliminating all mechanical pedestal devices.

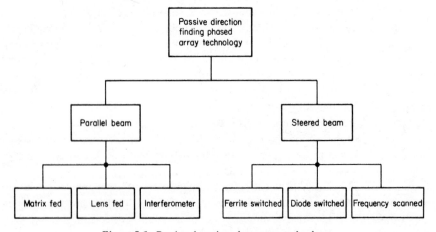

Figure 5-1. Passive detection phase array technology.

Any antenna can be considered to consist of a set of identical radiating elements, such as dipoles or slots, configured in a pattern such that under all conditions each element operates as each other. This includes the effect of mutual coupling between the elements and the effects common to all. For this given set there is an element pattern that is multipled by another pattern termed the array pattern (1) of the form:

$$g(u) = \sum_{n}^{N} \exp(jkx_n u) \int i_e(y) \exp(jkyu)dy \qquad (5\text{-}1)$$

$$= f(u)e(u) \qquad (5\text{-}2)$$

where

$$i_e = \text{the current density across the array}$$

$$x = \text{the location of the element } (x_n)$$

$$y = x - n_n$$

$$dy = dx$$

The two functions in Equation (5-2) are each radiation patterns; the first, $f(u)$, being the array factor, the second, $e(u)$, the element factor. This latter term is concerned with the excitation of each radiating element, which should be the same. The array factor is of prime importance since it establishes the pattern based upon the physical geometry of the radiating antenna structure. An array can be steered or scanned by series-feeding each element through an amplitude-weighted phase shift. This may be done in time by switching or by frequency scanning. To obtain fixed beamwidths that are always present, the elements of the array may be fed in parallel all of the time through a feed network that presents the appropriate phase differences to each radiating element. The steered-beam antenna can be made to exhibit array gain; the parallel beam antenna generally does not. We shall discuss the parallel array in this chapter as an array and in Chapter 6 as an interferometer. The frequency-scanned array, while of great importance in radar, finds little use in passive direction finding systems, since it does not generally exhibit the primary characteristic, wide instantaneous RF bandwidth, needed for high intercept probability.

5.2 PARALLEL BEAM FORMED ARRAYS

Since this type of antenna provides maximum coverage, it has become of great importance in passive ELINT and warning systems, where the complexity and size of the array and its associated circuitry can be tolerated. Most applications are found in naval systems. The planar array function can be obtained directly by a series of n parallel horn antennas as described in Chapter 4. In this case, the array is composed of the set of horn radiators, each of which exhibits gain in the direction of arrival perpendicular to the mounting plane. Amplitude comparison of the log-detected n channels will give the monopulse ratio to a high degree of accuracy and with a sensitivity determined by the horn gain over an isotropic antenna. The ratio can also be found by comparison to an omniazimuthal (360 degree field-of-view) or half-omniazimuthal (180 degree field-of-view) antenna for moderate accuracy and backlobe suppression as shown in Figure 4-17.

The advantage of this type of system is the ability to use conformal mount or geographical displacement of the antennas around a ship or plane. The disadvantage is the limited RF bandwidths of the horns. This is due to the fact that the gain of a horn antenna varies directly with the square of the frequency. The gain is usually chosen for a minimum gain at the low-frequency end, which results in high gain and an associated reduction of antenna beamwidth at the high-frequency end. The change in pattern shape, as a result, requires that frequency information be available to correct the DF look-up tables. An array of horns also suffers from spacing problems since high efficiency cannot be obtained with good beam crossovers unless the horns are closely spaced. It has also been determined that each horn should exhibit a sin x/x antenna pattern for maximum efficiency. This would lead to the conclusion that a uniform amplitude illumination be present, which is not practical. For these reasons, parallel horn systems become difficult to use although, simplistically, the ease with which paralleled amplitude detected channels can be compared is attractive. Double-ridged horns and frequency variable aperture area approaches have been used to extend horn bandwidths, but in general, paralleled horn feeds are confined to an octave, making any broadband (2–18 GHz) system relatively complex.

5.3 PLANAR BUTLER ARRAY

The matrix or hybrid network generally of the form developed by Butler (2) is another solution to arraying often used in planar arrays. This feed network typically takes the form of Figure 5-2, which is an n input, N output connection of a series of quadrature couplers, 180 degree hybrid combiners, and phase shifters interconnected to provide a progressive phase shift to each element of the planar array. The progressive phase shift concept derives from early work by Blass (9) and others. Essentially, a multiplicity of radiating elements will transmit a beam broadside to their plane if they are all in phase. If the phase between the elements changes progressively by $2\pi d/\lambda$, where d is the spacing between the elements, then the resulting pattern will point in the direction of the length of the line connecting all elements (endfire). It is then obvious that varying the phase progression between the two extremes can steer the resultant antenna beam.

The above case considers a feed network with one input and a multiplicity of outputs, each of which feeds a radiating element in some form of phase progression. If a parallel network of n inputs and N outputs is used, it is possible to generate parallel rather than scanned or sequential beams. Since these beams are generated by elements fed not in phase with one another, they can overlap without the undesirable coupling effects of the simple parallel horns described above, eliminating intercoupling losses and allowing attainment of an elemental sin x/x radiation pattern. Figure

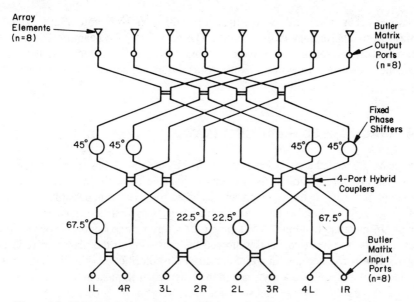

Figure 5-2. Eight-port Butler matrix using 90 degree hybrid couplers and phase shifters.

5-3 shows the beams developed by the network shown. Each beam is lettered left (L) or right (R) of boresight and is available at the bottom of the feed matrix for detection and comparison to an omniazimuthal antenna or the sum (from a detector) of all of the beams to form an amplitude monopulse ratio. It is also possible to measure the phase difference; however, the multiplicity of channels does not favor this approach. Reference (3) gives design equations and discusses three beam interconnections of the eight-element antenna. In the system described here, beam

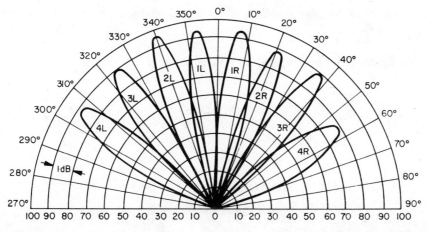

Figure 5-3. Simultaneous eight-beam radiation pattern from an antenna array fed by an eight-port Butler matrix.

crossover occurs at −4 dB, with adjacent beam isolation of about 20 dB. Antenna bandwidth is approximately one octave. The position of the beam peaks varies as a result of the presence of λ in the equations (Ref. 3 p. 263), requiring the use of frequency correction to assure that all dispersive phase shifts can be compensated for.

5.4   CIRCULAR BUTLER-FED ARRAY

The planar Butler-fed array is somewhat limited by the above considerations; however, it is also possible to utilize the Butler matrix feed to excite a circular array of radiating elements. If the elements are designed properly, it is possible to obtain a 3:1 or greater RF bandwidth. The circular array was first described in the early 1960s. In most applications, development of a series or sequential-switched beam was desired, and most concepts followed the reasoning of the planar array in this regard. It was realized that in a circular array the contribution of each element was not identical to the planar array since in the latter, each element is separately excited and can be easily expressed as such for each term in the general equation of the array factor.

In the circular array, the excitation can be considered to be a current distribution (4) $I(x)$ of many individual currents $I_n \exp(jnx)$, where n ranges between $-N$ and $+N$. The radiation pattern is

$$E(\theta,\phi) = \sum_{N=0}^{n=N-1} A_n e^{jn\phi} \tag{5-3}$$

where

N = the number of elements

A_n = a set of complex constants of current I_n multiplied by a Bessel function $J_n(x)$

When Equation 5-3 is compared for a linear array of N (elements),

$$E(\phi) = \sum_{n=0}^{n=N-1} A_n e^{jn\mu} \tag{5-4}$$

where

$E(\phi)$ = array factor

$A(n)$ = the nth amplitude excitation coefficients

K = d/λ

$$\mu \quad = 2\pi K \cos \phi$$

$$d \quad = \text{element spacing}$$

It may be seen that the circular array is similar to the linear array, differing mainly by the use of elemental current excitation in place of the amplitude or element excitation coefficients.

From this it follows that a circular array can be excited in a similar way compared to the linear array. Shelton (5) and Chadwick and Glass (6) describe networks to feed circular arrays in detail; however, it can be shown that the Butler matrix feed, described above, will provide parallel output patterns that are relatively insensitive to elevation θ, varying as the $\sin \theta$, and that the current excitation varies linearly across the elements with a phase difference of $(2\pi K)/N$. There will be N current modes corresponding to $K = 0 \pm 1, \ldots (N - 2)/2, N/2$. The total phase variation around the array is $2\pi K$ for the kth mode. By comparing the nonadjacent or plus and minus modes, for example, $K = +2$ to $n = -2$, an output phase variation of $n\phi$, in this case four times the arrival angle, will result. The Butler-fed array is therefore capable of high resolution (articulation), which we continually note to be the characteristic of phase monopulse systems that permit attainment of high-precision DF measurements with this technique.

The condition for high accuracy for a Butler-fed circular array dictates the use of more elements than output ports to increase phase resolution, which in turn necessitates that unused output ports be terminated. The Bessel term A_n in Equation (5-3) must be nonzero, requiring that the radius of the circular array be chosen to prevent cancellation of the radiator element contributions or that directional radiators be used to prevent mutual coupling effects. When the output ports of the Butler feed network are paired, narrow beams will not be obtained at each output port as was the case for the planar array $(n = N)$. Sets of phase differences are obtained between the pairs of ports that contain the angle of arrival, which, depending upon the pairing, is some multiple of the actual phase difference between radiating elements.

An excellent example of a Butler-fed circular array covering 2–18 GHz is a DF system designed by Anaren Microwave Inc., shown in Figure 5-4, a diagrammatic block diagram of the concept. The radiating structure consists of 32 horizontally polarized slot antennas each fed by a coaxial cable connected across the slot at its narrowest point. The slot tapers outward containing the horizontal E field, until the field radiates from each element as a horizontally polarized signal at the outer diameter. This is shown in Figure 5-5, a photograph of the printed circuit antenna element array mounted atop a foam spacer and the lower biconical reflector. The tips of the slots meet at the widest point, giving the appearance of a series of metallic petals; however, the slots, not the petals, radiate in the horizontal polarized direction. The antenna printed circuit card is mounted between two biconical reflectors purposefully to constrict the elevation angle (θ) to $+40 -10$ degrees from a zero degree horizon reference. The circular cutout behind each feed point is a high impedance matching transformer.

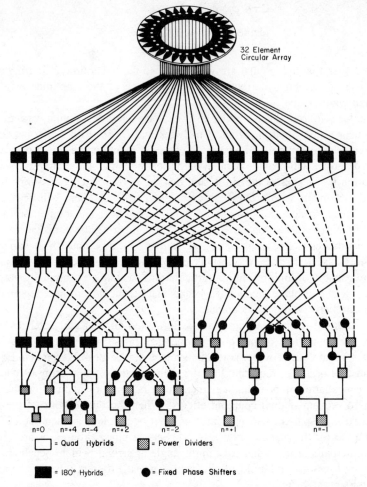

Figure 5-4. A 32-element Butler-fed circular array. Courtesy of Anaren Microwave Corp.

Each of the 32 coaxial cables is fed to an input port of a Butler matrix consisting of a series of hybrids, quadrature couplers, fixed phase shifters, and power dividers as indicated. This configuration is diagrammed in Figure 5-6, where it can be seen that the output ports are paired to provide $n = 0$, $n = \pm 1$, $n = \pm 2$, and $n = \pm 4$ beams. Only the reference $n = 0$ is compared to $n \times 1$, $n \times 2$, and $n \times 8$. This technique gives a phase variation of ϕ, 2ϕ, and 8ϕ for a physical input angle ϕ, thereby increasing the resolution of the system. The ambiguity of the high resolution $n \times 8$ port is compared to the $n = 0$ port giving high-phase sensitivity but with four ambiguities. These are resolved by phase comparison of the $n \times 2$ pair with the $n = 0$, which has two ambiguities that are finally resolved by the $n \times 1$ phase comparison to $n = 0$, which has no ambiguities. Since each phase angle is binary related, it is convenient to use this technique to develop an 8-bit digital word. Phase comparisons are made by broadband phase discriminators or correlators, another form of the class III monopulse system described in Chapter 2, Section 2.3. The correlator is described in detail in Chapter 10, Section 10.2.

Omniazimuthal output for an IFM or pulse detection channel can be obtained by amplitude detection of the $n = 0$ channel since this represents the energy summed from all elements of the array, and the system, in fact, makes use of this channel for that purpose.

The system shown has undergone extensive testing and is capable of providing 2 degree RMS accuracy over a wide range of operating characteristics. Figure 5-7 is a photograph of the 7.5–18.0 GHz array sited above the 2.0–7.5 GHz array, the latter covered by a polarizer that converts the received slant linearly polarized signal to horizontal polarization at the antenna element slots. The receiver attains a -60 dBm sensitivity in the low and -55 dBm in the high band. The overall operating accuracy when mounted in the radome (shown beside the antenna structure) and measured on a ship is between 1.78 and 2.6 degrees RMS, which are excellent results. This compares with a 1.5 degree RMS typical accuracy as reported by Anaren in anechoic chamber tests.

Figure 5-5. Anaren radiating structure. Courtesy of Anaren Microwave Corp.

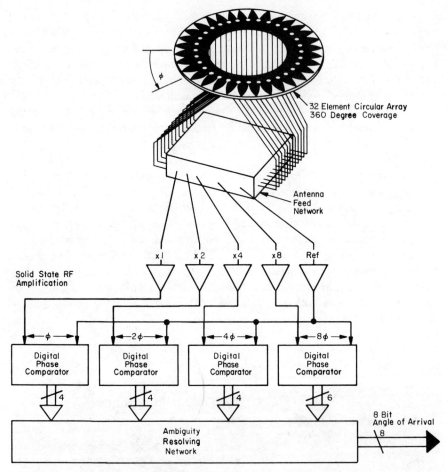

Figure 5-6. Configuration of a 32-element DF measurement system. Courtesy of Anaren Microwave Corp.

Figure 5-7. The 2–18 GHz Anaren circular array antenna system shown with polarizer and radome. Courtesy of Anaren Microwave Corp.

5.5 LENS-FED ARRAYS

The concept of a set of finite parallel beams formed by an array in real time has prompted adaptation of lens-fed technology. Although the Butler array, perhaps the most ideal feed for a multielement static DF system, is popular due to the ease with which the elements can be fed, it requires a fairly complex interconnection of components, all of which must be phase tracked in the configurations discussed. As the number of radiating elements increases, the feed network becomes more complex and expensive. Although the concept of lens-fed array technology was documented in the same time period as the Butler array, it was to take over 10 years until the state of the art of microwave printed circuitry and sophisticated computer design and measurements could advance to the point where practical lens array designs could be constructed.

The first lens array was patented by Gent (7) in England in 1958 for use as a feed for a radar scanner. It was a symmetrical or compound lens having as many feed elements or beamformer ports as element or antenna ports. This array was also known as the bootlace antenna. Figure 5-8 is a general diagram of an asymmetrical

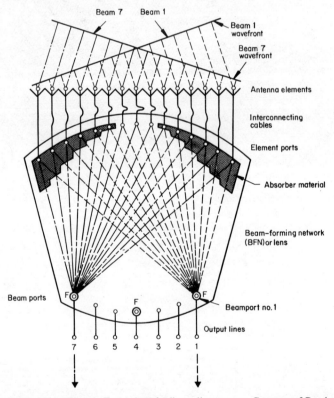

Figure 5-8. Parallel plate Rotman-Turner lens feeding a linear array. Courtesy of Raytheon Corp.

lens structure that has a different radius of curvature for the beamformer and antenna element ports. Seven beamformer ports and 15 element or antenna feed ports are shown. Each element port is connected to an antenna mounted in-line and connected to the lens by an interconnecting cable of a different physical length. Rotman and Turner (8) in 1963 described the operation of this type of lens in detail. Their original design used a parallel plate air dielectric pillbox structure for the lens element, which was capable of supporting the TEM mode of radiation. The beamforming ports consisted of radiating horns mounted at one end, which was a circular arc. RF probes at the element end fed a linear array as shown.

Conceptually, the Rotman lens works by summing or focusing N in-phase samples of a wavefront at a focal point. In Figure 5-8, consider the wavefront from beam 1. Each antenna element reading from left to right samples this wavefront at a slightly different time due to its angle of arrival with respect to the plane of the antenna elements. Each interconnecting cable between the antenna elements and the element ports is adjusted in length such that for three discrete angles of arrival, all signals add up in time at three focal points, shown as F at beam ports 1, 4, and 7. Tracing the left-most sample of wave front 1 to port 7 and comparing it to the right-most sample at the same port, it may be seen that the electrical path lengths are all equal due to the carefully adjusted lengths of the interconnecting cables. As a result, there will be a maximum at port 7 for beam 1. Similarly, if another wave front, beam 7, for example, arrives at a different angle, there will be a maximum at the focus at beam port 1. A third focus point is available at beam port 4. Any input wave front therefore will add up with a maximum at one port from 1 to 7, depending upon the angle of arrival of the corresponding wave front, and detection of the port amplitude will be a measure of the arrival angle. Although there are only three focal points, in practice the departure from exact focus at the other beamformer ports is not significant, in a good design, and all ports can be used to make accurate measurements.

Originally designed to be switch selected, the output beam ports may be parallel detected, resulting in simultaneous parallel formation of beams in a manner similar to a planar Butler-fed array. This is desirable for passive direction finding since the amplitude ratio of beams or the amplitude ratio of a beam to an omniazimuthal antenna will give a monopulse ratio. For retrodirective applications, the RF outputs of one lens array are not detected but are amplified by as many TWT amplifiers as beamformer ports. The TWT outputs are fed to a similar transmitting lens and antenna array. Since the array is linear, it follows the reciprocity theorem and a beam will be formed in the transmit array that will be the sum of all the TWT's power added in space and radiated in the same direction from which the original signal arrived. This retrodirectivity is the basis of many jammer designs, the advantage being that the total ERP is the sum of many lower-power devices, as opposed to a single high-power device. Redundancy is excellent since failure of one tube will not disable the system (see Section 4.7).

Since the addition at the focal points is due to equal phase path length for a wave front that arrives at a slightly different time at each antenna element, the Rotman lens can be considered to be a true time-delay multiple beamformer, similar to the Blass matrix (9). In Blass's design (Fig. 5-9), a linear array is fed by a series of transmission lines that cross another series effectively perpendicular to the first. At each junction point, there is a direction coupler to isolate the beams. The path lengths are adjusted to provide the same phase length from each antenna to each feed port, resulting in the same time delay. The patterns formed are shown in the figure and exist in space simultaneously. Since the Blass design (circa 1960) was executed in waveguide, it was dispersive (frequency sensitive) and troubled with delay-time formation that did not allow short pulses to sum instantaneously. The concept, however, was that of a time-delay array, of which the Rotman-Turner type is a prime example. The great advantage in a Rotman lens is the wide bandwidth and constant beam position attainable. This is due chiefly to the use of the TEM or coaxial modes of propagation for which the phase velocity is nondispersive. The Rotman lens is a true time-delay scanner that can be used either for receiving or transmitting.

The original Rotman-Turner version of the Gent bootlace lens was made up of a series of different-length coaxial cables. At one end the cables were terminated in a linear outer surface connected to an array of antenna elements. The other end of the set of different-length coaxial lines was connected to a parallel-plate section of maximum thickness terminated at the right by probes connected to the coaxial cables and at the left by horns or probes, constituting a set of F_N beam-former ports, of which F_1, G_1, and F_3 were in focus. Figure 5-10 shows the geometry of

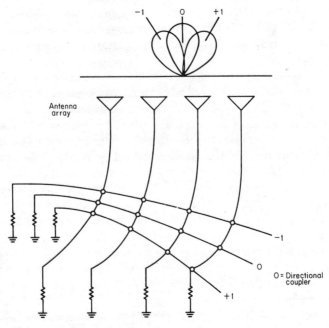

Figure 5-9. Blass equal length beam-former array.

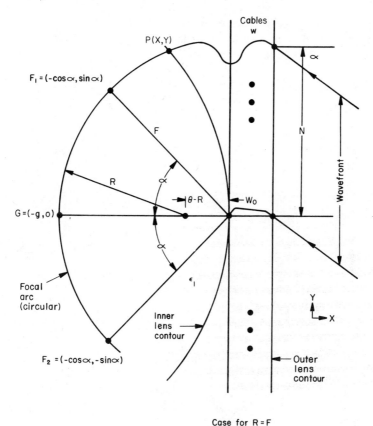

Case for R = F

Figure 5-10. Geometry of the Rotman lens.

the lens. These points were connected by a circular arc R. A signal from an input wave front arriving at an angle α at element N traveled a phase length w through cable, exited the cable at P_x, y, and traveled to the focus port through a ray path in the parallel plate section.

Archer et al. (10) improved the concept shown by introducing a dielectric material for air in the parallel plate section (constant ϵ_1) reducing the dimensions of this section by $1/\sqrt{\epsilon_1}$. This improvement also permitted the use of microstrip and stripline microwave printed circuit techniques to construct the feed section, eliminating the probes in favor of printed circuit launchers at the ports, with more determinable VSWRs.

The method of solution of the equations as outlined by Rotman is to choose α, normalize all values to F, the focal length, and determine the coordinates of the beam former and outer lens contour points. From this, the focal arc center can be computed. The distortion from the circular arc (phase error) is minimized by choosing $g = G/F = 1.137$ as a common value.

There are only three degrees of freedom: the cable length w and the X, Y coordinates of each point P of the inner lens contour. The equations for the lens design are tabulated as follows:

$$x = \frac{X}{F} \qquad w = \frac{W}{F} \qquad n = \frac{N}{F}$$

$$y = \frac{Y}{F} \qquad g = \frac{G}{F}$$

then

$$y = n(1 - w)$$

$$x = \frac{2w(1 - g) - b_o^2 n^2}{2g - a_o}$$

$$w = \frac{-b - \sqrt{b^2 - 4ac}}{2a}$$

$$a = \left[1 - n^2 - \left(\frac{g - 1}{g - a_o} \right)^2 \right]$$

$$b = \left[2g \left(\frac{g - 1}{g - a_o} \right) - \frac{(g - 1)}{(g - a_o)^2} b_o^2 n^2 + 2n^2 - 2g \right]$$

$$c = \left[\frac{g b_o^2 n^2}{g - a_o} - \frac{b_o^4 n^4}{4(g - a_o)^2} - n^2 \right]$$

where

$$a_o = \cos \alpha$$

$$b_o = \sin \alpha$$

R, the radius of the focal arc, lies at the center of at $R - G$:

$$R = \frac{(Fa_o - G)^2 + F^2 b_o^2}{2 (G - Fa_o)}$$

Since the above equations derived by Rotman used an air dielectric ($G = 1$), it is necessary to modify them for a finite dielectric by multiplication of X and Y by $1/\sqrt{\epsilon_1}$. The above equations may be readily solved for the general case with a personal computer using the program ROTMAN presented by Pozar (11) or for the symmetrical case as presented by Shelton (12).

The Rotman lens has assumed significant importance in naval electronic warfare (EW) systems as a result of the extensive work undertaken to reduce it to a small, reliable format by the Raytheon Corp. Goleta, CA. This has been accomplished by filling the parallel plate section with high dielectric materials with ϵ_r varying from 2.3 for Duroid Teflon fiberglass printed circuit material, to $\epsilon = 233$ for cadmium titanate material. Etched microstrip circuits on a barium tetratitanate ($\epsilon = 38$) ceramic substrate are also used. Figure 5-11 is a photograph of a Raytheon 66 element linear array fed by a 66 by 18 element beam-port lens. The azimuth patterns are shown in Figure 5-12a. There is about a 2 dB variation in pattern gain, which varies approximately as the cosine of the scan angle over a 120 degree field of view. Each pattern is 5 degrees wide at -3 dB, down broadening as the secant of the scan angle; backlobe levels are down by 24 dB. The elevation pattern (Fig. 5-12b) is approximately 10 degrees. The Rotman lens can also be used to feed moderately curved arc arrays. This is desirable for conformal applications. Figure 5-13 is a photograph of a 45 degree arc antenna covering 120 degrees that operates over a 2.5:1 frequency band. A four-sheet meander-line polarizer that converts circular and linear polarized signals to the linear polarization of the array is shown. This type of antenna offers improvement in gain roll-off at the extremes of angular coverage and can operate over wider RF bandwidths.

It is not feasible to use Rotman lens and linear or arc array to cover a full 180 or 360 degree azimuth angle since there is increasing rolloff in the contiguous beams as the angular sector is increased. A circular lens, however, known as the R-KR (13), may be used in a similar manner to develop semicircular contiguous beams. The lens diameter is K times the diameter of the array, where K is usually optimized at 1.9. Equal-length cables connect the antenna elements to the lens. Figure 5-14 is a photograph of a 17-element circularly polarized hemispherical azimuthal coverage antenna that operates over a 2.4:1 frequency band. Figure 5-15 shows the patterns obtained by this antenna. This type of array can be configured to cover 360 degrees

Figure 5-11. A 66-linear element array fed by a 66 × 18 beam-former lens. Courtesy of Raytheon Corp.

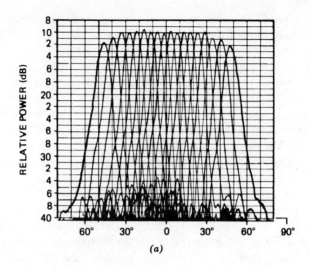

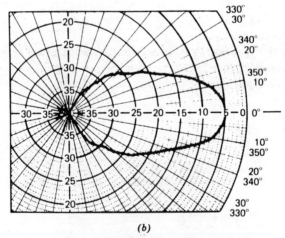

Figure 5-12. (*a*) Patterns of a linear 66 × 18 beam array. (*b*) Elevation (θ_H) pattern of 66 × 18 beam array. Courtesy of Raytheon Corp.

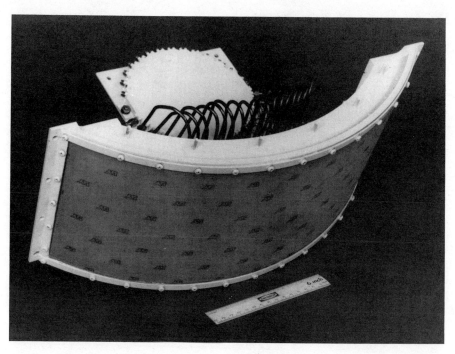

Figure 5-13. A 45 degree arc array fed by a stripline Rotman lens covering 120 degrees of azimuth.

Figure 5-14. The R-KR hemispherical multibeam lens-fed array. Courtesy of Raytheon Corp.

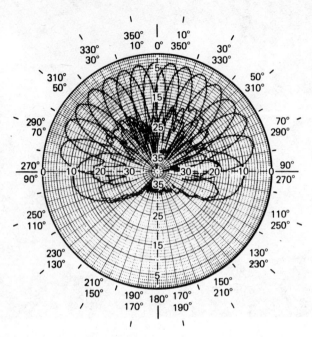

Figure 5-15. Patterns of the R-KR hemispherical multibeam lens-fed array. Courtesy of Raytheon Corp.

by interconnecting two lenses together with broadband 90 degree hybrids as shown in Figure 5-16.

The circular array features constant beamwidth patterns as a function of the scan angle, which is highly desirable. The maximum elevation angle coverage is

$$\theta_{max} = 202 \sqrt{\frac{\lambda}{D}}$$

where

$$D/\lambda = \text{the array diameter in wavelengths}$$

The antenna horizontal beamwidth is

$$\phi = \frac{59\lambda}{D}$$

This places a practical limitation on large diameter arrays since they will not have wide vertical beamwidth coverage. The Rotman lens as modified by Archer, Monser, and others at Raytheon has been chosen as the standard U.S. Navy electronic support and countermeasures (ESM/ECM) suite and is finding wide application in this role as well as in airborne and other applications. It has taken many years and much effort to convert the early Gent bootlace lens into a practical multibeam array.

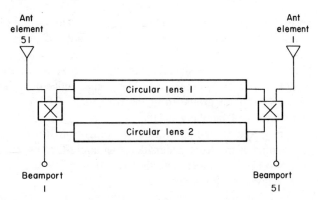

Figure 5-16. Omniazimuthal multibeam array utilizing two 90 degree hybrid interconnected R-KR lenses.

5.6 PARALLEL BEAM SEPTUM ANTENNA

In certain applications, it is necessary to develop direction-finding systems that can operate in limited space with elements that are not able to have sufficient aperture to attain high gain. These types of applications are generally found in aircraft installations and where space is at a premium. It is instructive to consider a rather unusual configuration provided by a corner reflector type of antenna, designed by the Elisra Corp., to cover an 8:1 frequency range at reasonable gain levels, with good DF accuracy.

Figure 5-17 is a photograph of a CD–L-band corner reflector antenna with its radome removed to show details. This antenna covers .5–2 GHz and provides three

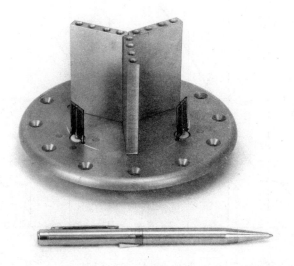

Figure 5-17. Three-element multibeam septum antenna. Courtesy of Elisra Corp.

unambiguous parallel beam outputs from three corner reflectors spaced 120 degrees apart. Each radiating element is a printed circuit broadband inverted-T section terminated at the far end. The center of the T is fed directly by a coaxial cable at the input port. The vertical members are fed from the center point and are terminated by resistive material to reduce reflections. The corner reflector is a three-piece triangular conducting septum grounded at the base plate. The 120 degree angle formed between the vertical members and the 90 degree angle formed by the septum and the baseplate tend to point the peak of the resulting beam downward; and since the antenna is mounted at the front underside of an aircraft, H-plane coverage (θ) is from the horizon to -45 degrees with the aircraft acting as a ground plane. A single cylindrical radome encloses the antenna.

Figures 5-18 and 5-19 are patterns taken at 700 and 1800 MHz at an elevation angle of $-22°$ below the horizon degrees for a vertically polarized source. At 700 MHz, crossover occurs at -3 dB points with a beamwidth of 220 degrees; at 1.8 GHz beam crossover occurs at 3.6 dB at 180 degree beamwidth. The gains of the

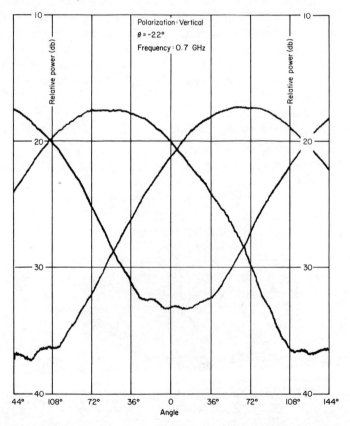

Figure 5-18. Parallel beam three-element septum antenna—0.7 GHz patterns.

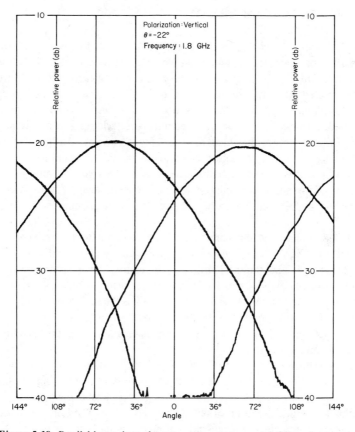

Figure 5-19. Parallel beam three-element septum antenna—1.8 GHz patterns.

antenna are listed in Table 5-1. The gain increases with increasing frequency, which can be compensated for by active RF amplification if required. DF determination is made by an amplitude comparison monopulse comparator. The absence of strong back lobes and the fairly predictable crossover levels permit a DF accuracy to be obtained to at least octant accuracy over the frequency range. The septum antenna approach can be extended to a greater number of elements and can be used with a Butler-feed network since the absence of backlobes indicates relatively good isolation between the elements.

Table 5-1. Septum Antenna Gain Characteristic

Frequency (GHz)	Gain ($\theta = -22°$) (dBi)
0.5	−8
1.0	+3
1.5	+5
2.0	+6

5.7 SWITCHED-BEAM ARRAYS

In many DF applications, it is not necessary or desirable to cover a full 360 degrees of azimuth. For example, in a situation where an ELINT aircraft is located at some distance from the expected targets, the basic direction-of-arrival of all desired signals is approximately known; what is needed generally is range and high precision. In geographic situations where the presence or turn on of radar emitters is to be determined, a sector DF system may be adequate. Although it is obvious that any of the multibeam approaches described thus far can be switched to selectively cover a given sector or beam in a planar sector, the advantage of attaining all of the array factor gain is not achieved if the unused ports of the switched multibeam must be terminated. From a transmit point of view, a switched-array that provides full array factor gain is highly desirable since it increases hard-to-achieve effective radiated power.

Switched-antenna systems are not commonly used with passive DF receivers due chiefly to the restricted instantaneous RF bandwidth usually associated with the feed network and the probability of detection loss resulting from the finite switching time. In the study of the Rotman lens system discussed above, switching can be used to describe which beam-former port has the maximum signal to determine the angle of arrival; however, parallel comparison is easier. In a multiple-horn system, switching can be used effectively to replace the rotating joint in a mechanically rotating antenna system.

In array technology, switching of phase and amplitude increments between various elements can be used to steer a beam. For missile work, switching of planar arrays with relatively narrow field-of-views is a highly practical method of accurate DF determination, since diode or varactor elements can be used due to the low-power application. Switch-steered arrays are extremely popular in radar work since great increases of effective radiated power are achieved. In these cases, locking ferrite switching devices add finite phase element increments to effect power summation of radiation from multielement arrays. Switched-phase lock injection accomplishes the same result from active element radiators in major systems.

In steered-array systems there are two basic types of effects that have a significant effect on their use for passive direction finding. Both effects are related to the instantaneous bandwidth that can be obtained. The first is the obvious limitation of the devices used to construct the array; that is, couplers, power dividers, phase shifters, and so on, each of which is imperfect to some extent. These effects are predictable. The second effect is due to the repeating nature of wavelength associated structures such as stubs, waveguides, and varying lengths. Only 2π or 360 degrees of phase change can be obtained in effect, and the rate at which it can be made useful depends upon distances between elements.

In actuality, a phased array would like to excite all radiating elements at the same *time*, with different phase and amplitudes at each of the elements to achieve directionality. The extent to which this can be accomplished varies with the configuration. A Rotman lens is a true time-delay phase shifter as are the Blass and other similar configurations. The characteristics of this class of device is beam pointing accuracy that is independent with frequency. Butler matrixes are not time-delay devices; however, if the *differential* phase shift is made directly proportional to frequency, the beam(s) will continuously point in one direction over wide bandwidths. The combination used for the circular array as described previously exhibits this property as a result of the choice of radiation and diameter and is a good example of which broadband circular arrays are often fed by the Butler matrix.

Steered-receive linear arrays have been developed utilizing phased-array techniques despite the above limitations. A planar steered-beam system is shown in Figure 5-20 and described in Ref. 14. In this case, a group or set of linearly aligned antennas is combined in series with a programmable phase shifter to apply a progressive time delay to each section permitting the formation of the two scanned beams. These two beams are combined in a $\Sigma-\Delta$ hybrid to form an amplitude signal pair that is subtracted to provide the angle of arrival ϕ as a function of the settings of the phase shifters, which are essentially scanned. The superheterodyne receiver is also tuned adding a second scanning probability to the detection process. It is claimed, however, that sensitivities of -70 dBm or greater can be achieved with this technique, exclusive of the antenna array gain. The DF accuracy is in the order of 1/30 of a beamwidth, the advantage of the system being the potential for full digital control and frequency correction.

There have been several attempts to design circular switched-phased arrays to replicate the scanning action of a rotating DF antenna. An early attempt by the

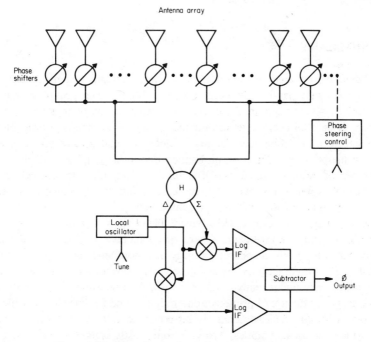

Figure 5-20. Azimuth scanning beam passive DF receiver.

U.S. Navy (15) resulted in the configuration shown in Figure 5-21. The concept was to make use of the circular array since, as described previously, the symmetry of the array is the same due to a constant relative phase shift between all elements. The antenna reportedly achieved 128 beam positions, with a crossover at -2 dB, a half-power beamwidth of 4–5 degrees, and a side-lobe level of 25 dB. Each step moved one element and was computer-controlled. The radiating elements were waveguide sectioned horns spaced at 0.65 wavelength in a cylinder of a 13.2 wavelength radius. All cables connecting the radiating horns were of equal electrical length. The phase shifters were PIN diode types and were followed by a programmable amplitude attenuator to vary the taper to control backlobes. The system achieved a 20% instantaneous bandwidth. In a later configuration, a R-2R lens similar to the type described for the Raytheon multibeam design, was used to eliminate the phase and amplitude switching by phase focusing through the action of the lens.

The above approaches, although somewhat complex, have been simplified by the availability of the frequency word from an instantaneous frequency measurement (IFM) or digital frequency encoder. As a result, future switch-scanned arrays may be developed to actually reduce the cost of the RF section of a DF receiver by the use of less expensive digital encoding storage and connection methods. The trend is clear for the transmit side, where high effective radiated power is obtained more inexpensively by a multiplicity of amplifiers and radiating elements, with the added advantage of redundancy and reliability. The need for higher sensitivities can be best achieved by narrower steered DF beam antennas, where high instantaneous intercept probability is not required.

5.8 SUMMARY

This chapter has covered the important phased array antenna systems used in modern electronic warfare passive direction finding systems. The groupings have been characteristically divided into the way the detection problem stands—multiple coverage or parallel beam antennas for maximum spatial intercept probability versus switched-beam techniques for higher accuracy and greater effective antenna gain—at the price of a detection probability due to the required scanning process. Both types of systems are in use. The third system, the interferometer, will be detailed in Chapter 6; however, it may be seen that this approach is, in effect, a form of both types of the antennas presented here.

It is important to note that phased-array DF systems lend themselves readily to conformal antenna arrangements where the vehicle is enclosed by the antenna system. Since the performance of an array has more antenna elements, it is more tolerant of perturbations than say a four-element system and is therefore capable of better accuracy under many conditions. It is also possible to reduce the cost of receive arrays by the use of printed circuit radiators and feed networks, which allow the array, with all of its apparent circuit complexity, to be reduced to relatively easy-to-manufacture assemblies. Arrays become costly as they are required to handle power, the receiver arrays becoming more practical as feed networks improve.

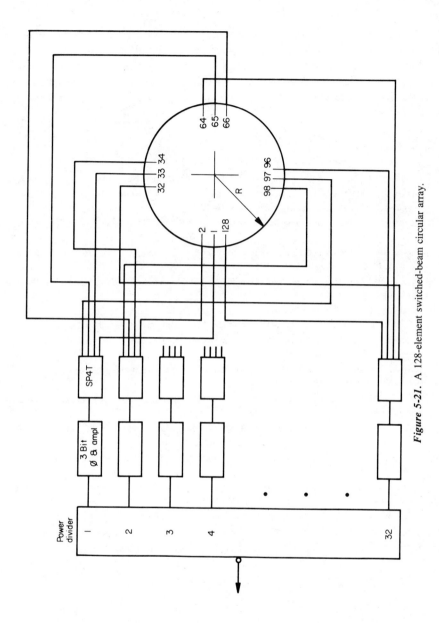

Figure 5-21. A 128-element switched-beam circular array.

For these reasons, the topic of DF antenna arrays has been given more attention than current usage would imply. As higher frequency millimeter wave technology evolves, the phased array is certain to play an important part in passive direction finding for missiles, remotely piloted vehicles, and spaceborne vehicles due to the smaller sizes achievable.

REFERENCES

1. Steinberg, B. D., *Principles of Aperture & Array System Design*, New York: Wiley, 1976, p. 76.
2. Butler, J., and R. Lowe, "Beam Forming Matrix Simplifies Design of Electronically Scanned Antennas," *Electronic Design 9*, 1961, pp. 170–173.
3. Hansen, R. C., Ed., *Microwave Scanning Antennas*, New York: Academic Press, 1966, Vol. III, p. 260.
4. Sheleg, B., "A Matrix-Fed Circular Array for Continuous Scanning," *Proc. IEEE*, Vol. 56, No. 11, Nov. 68.
5. Shelton, P., "Application of Frequency Scanning to Circular Arrays," IRE WESCON Conv. Record Pt. 1, 1960, p. 83.
6. Chadwick, G., and J. Glass, "Investigation of a Multiple Beam Scanning Circular Array," Report 1—USAF Cambridge Research Lab, Cambridge, MA, Contract AF 19/628/367, Dec. 1962.
7. Gent, H., "The Bootlace Aerial," *Royal Radar Establishment Journal*, England, Oct. 1957, pp. 47–57.
8. Rotman, W., and R. Turner, "Wide-Angle Microwave Lens for Line Source Applications," *IEEE AP-T*, Nov. 1963, p. 623.
9. Blass, J., "Multi-Directional Antenna—A New Approach to Stacked Beams," *IRE National Convention Record*, 1960, Part 1, pp. 48–50.
10. Archer, D., R. Prickett, and C. Hartwig, "Multi-Beam Array Antenna," U.S. Patent 3,761,936, Sept. 25, 1973.
11. Pozar, D., "Antenna Design Using Personal Computers," Dedham, MA: Artech House, 1985, p. 94.
12. Shelton, J. P., "Focusing Characteristics of Symmetrically Configured Bootlace Lenses," NRL Memorandum Report 3483, April 1977, ADA 039843.
13. Thies, W., Jr., "Omnidirectional Multibeam Antenna," U.S. Patent 3,754,270, May 1974.
14. Simpson, M., "High ERP Phased Array ECM Systems," *Journal of Electronic Defense*, Mar. 1982.
15. Boyns, J., "Step Scanned Circular-Array Antenna," Phased Array Conference, 1977.

Chapter Six ─────────────────────────

Interferometer
DF Techniques

Interferometers can be considered specific cases of array antennas. The linear array in which all antenna elements center lie in a straight line at equal spacings with beam forming done by phase shift networks is an example. In an interferometer, the elements can also lie in a straight line but are usually at different spacings to obtain time-of-arrival relationships that can be transposed into measurable phase differences for determination of angle-of-arrival information. The concept also extends to the circular array, where interferometer techniques can be applied to also cover 360 degrees of azimuth from an antenna mounted at one point.

In Chapter 2, Section 2.21, the simple interferometer antenna system was described as a DF measurement technique based upon the difference of the time of arrival of a signal detected by two identical collocated antennas in space separated by a finite base line. It was shown that the output of the antennas differed in phase from each other in proportion to the extra time it took a plane wave signal to travel a greater distance to the further antenna. Since a signal would be detected at the same time if it were equally displaced from both antennas, the phase difference would be zero, creating a phase difference null along the boresight for this condition. The measurement of the time difference or phase, whenever it is greater or less than zero, therefore, can give the incoming angle with respect to boresight.

A DF system consisting of two monopulses in space is 180 degrees ambiguous since it is not clear from which half of the hemisphere the signal originated. Practical interferometer systems solve this problem by using another system such as an amplitude monopulse DF to select the proper direction, or use quadrant arrays of antennas shielded from each other, or a circular array of monopoles to unambiguously cover a 360 degree field of view. In the first technique, ambiguity resolution is obtained by shielding, which is accomplished by utilizing an artificial ground plane, such as the side of a ship or plane, to absorb or reflect the back wave. The last method resolves the ambiguity by measuring the relative differences in phases between all the radiating elements around a circle, essentially resolving ambiguous responses.

Wide frequency coverage is generally desired in interferometer systems used for passive detection, necessitating that more than one set of antennas and more than one fixed baseline between the antennas be used, since the solution of the angle-of-arrival equations for a single baseline system is not single valued. Interferometers of this type are called multiple baseline types and generally consist of a reference antenna in the center of two or more other antennas spaced in line and at different distances from it. It is possible to use many baselines and antennas to obtain wide frequency coverage. By relating the baselines in a known manner or by using the known RF frequency in the processing, all DF ambiguities can be resolved (1). In another technique the phase spacing can be made proportional to frequency by pointing two conical antennas toward each other. In this case the phase null ambiguity is resolved since the spacing optimizes itself for each frequency. The wide bandwidth of conical spirals makes this a useful technique (2). In this chapter we shall consider several forms of the multiple baseline interferometer with both sectorized linear and circularly dispersed elements. Examples of the system configurations of both types will be given to show typical applications for dispersed and single-point antennas.

6.1 MATHEMATICS OF INTERFEROMETRY

At the outset, it is important to understand that the antenna part of interferometer systems is almost always a phase monopulse configuration, permitting the processing techniques outlined in Chapter 2 to apply. The practical problem of measurement of phase, however, often requires that combinational phase and amplitude systems be used. An example of this is the resolution of previously mentioned ambiguity: By detecting the antenna amplitude as well as phase responses and by comparing sectors in an amplitude monopulse system, it is possible to eliminate the multiple directional null ambiguities of the phase interferometer that occur when frequencies having wavelengths less than the minimum unambiguous spacing ($\lambda/2$) of an antenna pair are present.

In Figure 6-1 ,the single baseline two-antenna interferometer has been drawn to show operation in the first quadrant. The voltage received by antenna 2 may be

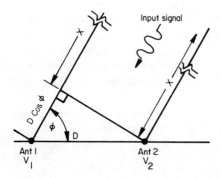

Figure 6-1. Single baseline interferometer.

expressed in exponential form as follows:

$$V_2 = V \exp\left(j\omega t - \frac{2\pi}{\lambda} X \right) \qquad (6\text{-}1)$$

where

V = the initial transmitted signal amplitude

X = the distance traveled

$2\pi/\lambda$ = free space propagation constant

and in a like manner at antenna 1:

$$V_1 = V \exp\left(j\omega t - \frac{2\pi}{\lambda} X + D \cos \phi \right) \qquad (6\text{-}2)$$

where the term $D \cos \phi$ represents the additional path length to antenna 1 as referenced to antenna 2. Since we are interested in relative phase differences between the two received voltages, V_1 and V_2, we can assume $(2\pi/\lambda)X$ to be reference zero at antenna 2; then,

$$V_2 = e^{j\omega t} \qquad (6\text{-}3)$$

and

$$V_1 = \exp\left(j\omega t + \frac{2\pi}{\lambda} D \cos \phi \right) \qquad (6\text{-}4)$$

Taking the natural log (ln) of both sides of the equation and subtracting to obtain the difference voltages yields

$$\ln V_2 - \ln V_1 = \ln \frac{V_2}{V_1} = j\omega t - jwt + jD \cos \phi \qquad (6\text{-}5)$$

$$= j \frac{2\pi}{\lambda} D \cos \phi \qquad (6\text{-}6)$$

Let Ψ be defined as this difference or the real part of Equation (6-5); then,

$$\Delta \Psi = \frac{2\pi}{\lambda} D \cos \phi \qquad (6\text{-}7)$$

Taking the difference of the two natural log voltages effectively forms the ratio $\ln V_2/V_1$, which makes this interferometer a phase monopulse system; the $\Delta \Psi$ term has been normalized with respect to input signal variation and is the monopulse ratio.

Substituting

$$\Delta\Psi = \frac{c}{f}, \qquad \text{where } c = 3 \times 10^{10} \text{ cm/sec}$$

and solving Equation (6-7) for cos ϕ in terms of frequency yields

$$\cos \phi = \frac{30\Delta\Psi}{2\pi Df} \qquad (6-8)$$

where

$\Delta\Psi$ = the phase difference in radians (2π radians = 360 degrees)

f = the frequency of the intercept expressed in gigahertz per second

D = the spacing measured in centimeters

ϕ = the angle of arrival

or

$$\cos \phi = \frac{\Delta\Psi}{12Df} \qquad \text{(in degrees)} \qquad (6-9)$$

Since $\Delta\Psi$ repeats itself every 360 degrees for all higher frequencies where the incoming wavelength is less than D, for the cos ϕ in Figure 6-1 to be unambiguous, the relative phase shift $\Delta\Psi$ must be

$$0 < \Delta\Psi < 360 \text{ degrees} \qquad (6-10)$$

For the general case, then

$$\cos \phi = \frac{\Delta\Psi + 360K}{12fD} \qquad (6-11)$$

where K is any positive integer that is the solution number corresponding to higher frequencies. Since the cosine is always less than 1, K is limited to only those values that make cos $\phi \leq 1$ in Equation (6-11).

The two-antenna interferometer is therefore a limited bandwidth device since if the RF bandwidth is extended higher, a greater number of K solutions will result and it will be impossible to know which K is correct.

This can be shown by rearranging terms in (6-11):

$$\cos \phi = \frac{\Delta\Psi + 2\pi K}{12fD} \qquad (6-12)$$

Substituting $f = B/\lambda$, where B contains all constants, and dividing by 2 gives

$$B \frac{D}{\lambda} \cos \phi = \frac{\Delta \Psi}{2\pi} + K \qquad (6\text{-}13)$$

Figure 6-2 is a plot of ϕ, the angle of arrival (as measured from the horizon 0–90 degrees in the first quadrant) as a function of $(\Delta \Psi / 2\pi) + K$.

From the plot it may be seen that there will be no ambiguities for $D/\lambda = \frac{1}{2}$. If we consider any other spacing to wavelength ratio such as $D/\lambda = 4$, for example, for a change from 10 to 80 degrees, there will be a change in $(\Delta \Psi / 2\pi) + K$ from 3.6 to .5 or 3.1 degrees of rotation, creating three ambiguities. The single baseline system would therefore not be single-valued over this span.

Consider, however, the articulation, which for an interferometer can be defined here as rate of change of angle of phase for angle of arrival. If we define the slope $m_{0.5}$ to be that of the $D/\lambda = \frac{1}{2}$, then

$$\frac{.5}{90 \text{ degrees}} \times \frac{180 \text{ degrees}}{\pi(\text{rad})} = .32 \text{ degree/degree}$$

For $D/\lambda = 4$, the slope m_4 for this case is

$$\frac{4}{90 \text{ degrees}} \times \frac{180}{\pi} = 2.55 \text{ degree/degree}$$

which is eight times better than $D = \frac{1}{2}$ $(4/.5)$ as expected; therefore, it is possible to conclude that *the interferometer DF technique can achieve very high resolution by a choice of large D/λ ratios at the expense of solving the ambiguity problem*

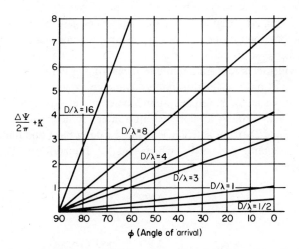

Figure 6-2. Interferometer ambiguity $(\Delta \Psi / 2\pi + K)$ as a function of the angle of arrival and spacing to wavelength of a single baseline interferometer.

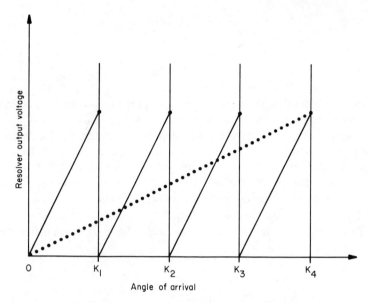

Figure 6-3. Ambiguity resolution in a multiple baseline interferometer.

that will result. This is the basis for the use of this technique in high resolution DF systems.

One method of resolving the ambiguities over a limited frequency range is to limit the field-of-view of the antenna. Another method is to use a multiple baseline. Figure 6-3 shows the latter concept: An ambiguous baseline pair gives rise to ambiguities K_1, K_2, K_3, and K_4 (solid line), which can be resolved by another baseline pair (dotted line) having no ambiguities, the second pair being more closely spaced. The fine angular measurement is best measured by the widest space pair since it has the greatest articulation (phase change per degree); however, the most significant number for the measurement, the unambiguous one, is made by the channel with the least articulation, requiring a better signal-to-noise ratio in this channel to provide accuracy. An amplitude-measuring system is often used here to provide this resolution if a signal-to-noise advantage exists.

Figure 6-4 shows a dual baseline system where antenna 1 is spaced at a distance D_1 from antenna 2 forming a short baseline interferometer, while antenna 3 is spaced at a distance D_2 from antenna 1 forming a long base-line interferometer.

In the first quadrant, V_3 lags V_2, which lags V_1. Equation (6-11) rewritten for this case becomes

$$\cos \phi_1 = \frac{\Delta \Psi_1 + 360 K_1}{12 f D_1} \tag{6-14}$$

$$\cos \phi_2 = \frac{\Delta \Psi_2 + 360 K_2}{12 f D_2} \tag{6-15}$$

where

$$D_1 \neq D_2$$

For a given range of frequencies, the values of K_1 and K_2 can be computed by setting cos ϕ_1 and cos ϕ_2 equal to unity and solving for K. For example, using Equation (6-14)

$$\cos \phi = \frac{\Delta \Psi_1 + 360 K_1}{12 f D_1} \leq 1$$

Then

$$\Delta \Psi + 360 K_1 \leq 12 f D_1$$

but

$$\Delta \Psi = 0$$

$$360 K_1 = 12 f D_1$$

$$K_1 = \frac{f D_1}{30}$$

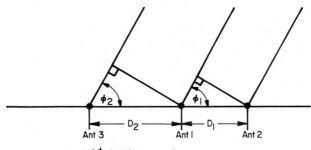

a. $0 \leq \phi \leq 90$ (Quadrant I)

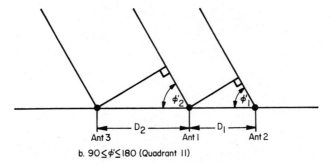

b. $90 \leq \phi' \leq 180$ (Quadrant II)

Figure 6-4. Dual baseline interferometer. (*a*) Quadrant I, where $0 \leq \phi \leq 90$ degrees. (*b*) Quadrant II, where $90 \leq \phi \leq 180$ degrees.

K_1 can have multiple values for higher frequencies as before (see D/λ curves in Fig. 6-2). The number of ambiguities will be an integer ranging from zero, the ideal case, to any integer less than the solution.

In quadrant II the angle of arrival becomes a lead angle; that is, V_3 leads V_1, which leads V_2. This is shown in Figure 6-4b.

At 90 degrees or boresight there is no differential angle and the angle-of-arrival reverses from lag to lead. Since most real-time measuring systems measure the lag angle, some adjustment to the equations is required. Defining quadrant II values as primed, and measuring them as shown, Equation (6-9) becomes

$$\cos \varphi^1 = -\frac{\Delta \phi'}{12Df} \qquad (6\text{-}16)$$

$\Delta \Psi'$ is a lead angle now and is therefore negative with respect to the measurement system. Figure 6-5 shows the quadrant relationships. The signal arriving at ϕ^1 in quadrant II would have the same value as $\cos(\pi - \phi)$ for an equivalent in quadrant I where

$$\pi/2 \quad < \phi^1 \leq \pi$$

$$\cos \phi^1 = \cos(\pi - \phi)$$

$$= \cos \pi \cos \phi + \sin \pi \sin \phi$$

$$= -\cos \phi$$

$$\cos \phi = \frac{\Delta \Psi}{12fD} \qquad (6\text{-}17)$$

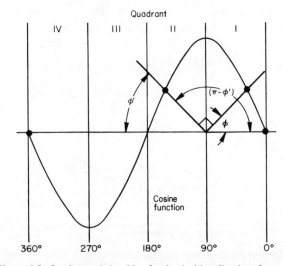

Figure 6-5. Quadrant relationships for the dual baseline interferometer.

Since only lag angles are measured, Ψ^1 would be advanced 360 degrees as follows

$$\Delta\Psi^1 \;=\; 360 \;+\; \Delta\Psi \tag{6-18}$$

Substituting Equations (6-18) into (6-17) after solving (6-18) for $\Delta\Psi$ gives

$$\cos\phi \;=\; \frac{(\Delta\Psi^1 \;-\; 360)}{12fD}$$

Since only angles from 0 to 360 degrees can be measured for K solutions,

$$\cos\phi \;=\; \frac{(\Delta\Psi^1 \;-\; 360K) \;-\; 360}{12fD}$$

The angles of arrival ϕ_1 and ϕ_2 for quadrant II may therefore be written

$$\cos\phi_1 \;=\; \frac{\Psi_1 \;-\; 360(K_1 \;+\; 1)}{12fD_1} \tag{6-19}$$

$$\cos\phi_2 \;=\; \frac{\Psi_2 \;-\; 360(K_2 \;+\; 1)}{12fD_2} \tag{6-20}$$

where $\Delta\Psi_1 = \Psi_1$ and $\Delta\Psi_2 = \Psi_2$ by definition, the true measured phase angle differences at the antennas.

6.2   SOLUTION OF THE INTERFEROMETER EQUATIONS

In the above section, the equations for the cosine of the angle-of-arrival were derived for a dual baseline three-antenna interferometer. It was shown that for all frequencies where the wavelength of the incoming signal was less than twice the distance between any two antennas, multiple solutions for the cosine would appear requiring the use of another set of antennas at a different distance to select the proper Kth solution. This establishes the maximum bound for no ambiguities of $K = 2D/\lambda$. It was also shown that the rate of change of the phase angle increased as more wavelengths could be fitted within a given distance, which is also the condition for more ambiguities.

The ideal dual base-line interferometer uses the shorter distance to resolve the ambiguities and the longer baseline to give maximum resolution. By proper choice of the ratios of the distance of the three antennas, ambiguities can be resolved by the use of common readings from both channels.

The system that makes use of the equations derived above makes the assumption that the cosines of the angle-of-arrival are all equal since the baseline distances are small with respect to the distances from the received emitter, that is,

$$\cos \phi_1 = \cos \phi_2$$

Since this is the case the equations for quadrant I, Equations (6-14) and (6-15), can be set equal, or

$$\frac{\Delta \Psi_1 + 360 K_1}{12 f D_1} = \frac{\Delta \Psi_2 + 360 K_2}{12 f D_2}$$

$$\frac{\Delta \Psi_1 + 360 K_1}{\Delta \Psi_2 + 360 K_2} = \frac{D_1}{D_2} \triangleq \frac{1}{R} \tag{6-21}$$

where R is defined as the spacing ratio D_2/D_1.

This shows that Equation (6-21) is independent of frequency and that the only restriction on the dual (or multiple) baseline interferometer is the selection of R such that an unambiguous solution can be found. Although the frequency is not required for the relative difference in phase in the antenna voltages, as a practical matter for a wide-frequency bandwidth, frequency information is required to permit the calculation of ϕ in each of the $\cos \phi$ baseline equations. This is necessary to determine the values of K_1 and K_2 necessary to eliminate ambiguities. Also, frequency information is often used in look-up tables for correction of angle-of-arrival data that may be frequency dependent. Since the interferometer described here requires phase matching up to the measurement of antenna-to-antenna phase, dispersion due to frequency can be removed by calibration of the phase-to-angle look-up table in the angle-conversion computer.

Equation (6-21) can be rearranged as

$$R \Delta \Psi_1 - \Delta \Psi_2 + 360 (R K_1 - K_2) = 0 \tag{6-22}$$

and grouped with the quadrant 2 equations using the above techniques as follows:

$$R \Delta \Psi_1 - \Delta \Psi_2 - 360 [R(K_1 + 1) - K_2 - 1] = 0 \tag{6-23}$$

Equations (6-22) and (6-23) can be solved for $\Delta \Psi_1$ and $\Delta \Psi_2$. These are idealized equations and must be corrected for errors in quantization or conversion of phase to a digital number. The solution of Equations (6-14) and (6-15) in quadrant 1 and Equations (6-19) and (6-20) in quadrant 2, utilizing frequency information, will give a family of numbers for K_1 and K_2 that relate to each other in a deductive way in proportion to the method and resolution of the quantization. This can be done by setting Equations (6-22) and (6-23) equal to two separate error constants that account for nonideal measurement of $\Delta \Psi_1$ and $\Delta \Psi_2$. If the measurement of phase is done by digital quantization, then the phase error constants will approach the least significant digital phase increment plus or minus other phase errors in the system. Since the R factor, which is the ratio of the longer to shorter baseline (in our convention), weighs $\Delta \Psi_1$, in both of the above equations, greater deviations in the measurement of $\Delta \Psi_1$ can be tolerated.

The solution method is to let each $\Delta\Psi$ be represented by the sum of two phase angles $\Psi + \Psi'$, the first being the quantized value, the primed being the independent system error. Equations (6-22) and (6-23) can then be expressed in terms of the quantifying increment $360/Q$, where Q is the minimum resolution of the encoding system. By substituting the known values for R and Q and solving (6-22) and (6-23) for the measured phase differences $\Psi_1 - \Psi_2$ in terms of K, the system errors, and Ψ_1' and Ψ_2', it is possible to develop a family of K numbers, which when compared to the quantum number Q at various frequencies can provide a unique solution to the interferometer ambiguity resolution. The special case of $\Delta\Psi_1 = \Delta\Psi_2 = 0$ occurring at system boresight is recognized by the presence of signals in both channels at the same frequency while at differential phase zero.

The determination of the angle-of-arrival by interferometer methods will be further detailed in Chapter 9 discussing accuracy and signal-to-noise ratios; however, it is useful to again point out that the base accuracy and signal-to-noise ratio are achieved by the longest baseline interferometer, which has the highest probability of error due to multiple ambiguities yet provides the best articulation. The short baseline interferometer gives the poorest signal-to-noise performance but can be made unambiguous. In a quantized system, the short baseline yields the most significant digit of the DF measurement and often determines the sensitivity of the system.

The mathematical details of a two-quadrant planar (in-line) interferometer system have been presented above to give the reader a basic idea of how interferometers work and the significance of the maximum usable high frequency and the ambiguities that result. The derivation method was chosen to show how the details can be worked out since most texts assume that once the concept is presented, the reader can easily proceed to a solution, which is not always the case.

In practical systems, phase-only interferometers suffer from accuracy problems. Although they have the potential for very high accuracy, it is coupled with finite probability of incorrect ambiguity resolution, called gross error. Although frequency-independent systems can be built, knowledge of the intercept frequency is an important tool in reduction of the above error probability. Reference 3 gives curves of probability of ambiguity as a function of the bounds on the ratio of the separations between the antennas, assuming that the separations are not equal ($D_1 \neq D_2$ in Fig. 6-4). It can be concluded that there is a higher possibility for error. For this reason most interferometers used in passive DF work are designed to be planar multiple antenna sections, conformally dispersed about a host vehicle, and used in conjunction with amplitude comparison techniques. This is done by comparing the amplitude of each section in a manner similar to that done for a four-spiral DF system as an ambiguity resolving method. When conformal mounting can be replaced by single-point mounting, a cylindrical or multielement Butler-fed array, or circular interferometer, is used.

6.3 MULTIPLE-APERTURE SYSTEMS

It is possible to utilize related multiple antenna apertures and multiple baselines in known ratios to make the determination of the angle of arrival unambiguous over

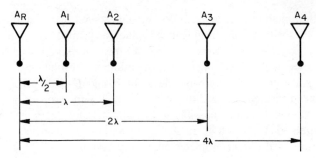

Figure 6-6. Unambiguous multiple baseline interferometer system with binary relationships between spacings.

a wide frequency range. Reference 4 describes a binary-beam interferometer system that divides each aperture base line by a factor of 2 as shown in Figure 6-6. For a binary variation of 2, between pairs the system is unambiguous from boresight to the ± 90 degree maximum field-of-view ($K_1 = K_2 = 0$) since the cosine of the angle of arrival is uniquely determinable. The first antenna pair forms a $\lambda/2$ baseline with greater slope (articulation) than the next but with ambiguities which are resolved by the next pair. The process continues up to the longest baseline (shown in the figure).

When a signal is received, the entire DF word is defined. The interferometer is used to remove ambiguities of the 2λ pair, which removes ambiguities of the λ pair, and so on. The resolution can be as great as the number of baselines used. Frequency information can be obtained to provide the means to determine the K numbers as well as provide intercept frequency.

In a multiple-aperture or baseline system there are two basic choices to be made: a harmonic binary-related system as described above or a nonharmonic-related spaced baseline array. As shown, the harmonic or, in this case, binary system provides a completely unambiguous output of DF up to the highest operating frequency by the antenna pair relationship. In the nonharmonic system, ambiguities will be present at many frequencies; however, knowledge of the spacing ratios and the frequency of the incoming signal permits computation by the use of digital look-up tables and logic methods. Any errors or mistrack in the RF receiving system can be incorporated into the correction.

The trend in interferometers is toward the nonharmonic type since a greater RF bandwidth coverage can be provided in contrast to the original concept of the binary-type system. This is due to the improvement in instantaneous frequency measurement techniques and the fact that, in actuality, both types of interferometers need frequency corrections and, of necessity, utilize superheterodyne techniques to limit the instantaneous frequency bandwidth. A binary system is limited on its use at the high-end frequency range to the half-wavelength spacing and at the lowest frequency by physical dimension. In practice, it is claimed (5) that 2–18 GHz coverage can be provided with nonharmonic spacing, related antennas with 0.1–3 degrees RMS accuracies.

Antennas used for the above systems are generally planar circularly polarized archimedian spiral antennas, which are mounted in the baseline surface and rotated, or clocked, to obtain the best phase matching. Since the system is a phase monopulse receiver, amplitude balance is less important, and since the phase variation doubles for each binary baseline increase, phase matching becomes less critical for the longest baseline channel.

6.4   LINEAR AND CIRCULAR INTERFEROMETER ARRAYS

An interferometer system making use of many of the concepts presented here is shown in Figure 6-7, a generic four-quadrant DF system patterned after Ref. 6. This system utilizes four quadrant superheterodyne receivers geographically dispersed around the host vehicle, each providing coverage of a 90 degree sector. A

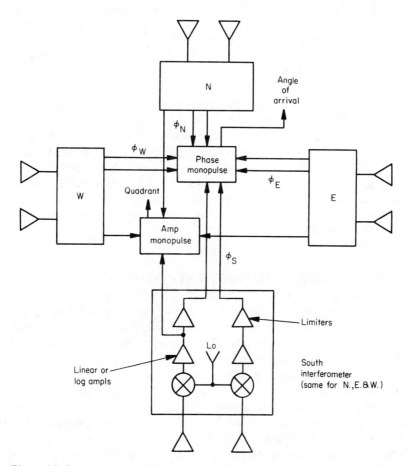

Figure 6-7. Phase monopulse DF receiver with amplitude monopulse ambiguity resolution.

central processing unit contains a common local oscillator, which is fed to each set of mixers in each quadrant receiver, ensuring that every DF channel will be synchronously tuned. An instantaneous frequency receiver measures the frequency of the local oscillators in the narrow bandwidth superheterodyne to determine the frequency for K selection of the interferometer. Signal frequency is also measured by the IFM to determine and sort signals. An amplitude monopulse comparator compares the amplitude of the detector log video amplifier (DLVA) output taken from each quadrant receiver to eliminate back lobes and give a coarse DF angle.

A single receiver that is identical to the other three in operation is shown in detail. Each incoming signal is received by a multiple baseline interferometer system, consisting of a set of circularly polarized spiral antennas connected to a narrow bandwidth sweeping superheterodyne receiver. There are as many channels as antennas, each matched in phase to the extent necessary. (The unambiguous $\lambda/2$ channel requires the closest phase match.) The angle-of-arrival phase and amplitude monopulse circuits operate on each phase-matched channel and by knowledge of the local oscillator and signal frequencies calculate the complete angle-of-arrival, resolving the K ambiguity numbers. The phase monopulse comparator receives amplitude limited signals and operates essentially as a class II processor. Outputs from the DF system are combined with pulse-time information for signal analysis processing in the computer.

A singular advantage of a quadrant-type system of this type is the clear angle of view that can be obtained by dispersement of the antenna and receivers around a ship or aircraft. Since the only RF signal distributed over a long length is the local oscillator, maximum sensitivity can be obtained. Bearing accuracy is excellent, and the system can be made to cover a large frequency range, typically 2–18 GHz. The cost of this type of equipment is high, however, due to the plurality of expensive superheterodyne receivers and processing channels. Intercept probability due to the reduced instantaneous RF bandwidth is moderate, being acceptable in high-density environments where the superheterodyne scan probability works in favor of unburdening the processing system.

6.5   CIRCULAR INTERFEROMETER SYSTEMS

The concept of the interferometer as a linear array can be extended to include the circular array. The linear baseline array, which we discussed above, is usually configured in a conformal geographically dispersed system, where each quadrant operates essentially as a separate azimuth elevation sensor.

Figure 6-8 is a diagram of a four-element 2–18 GHz interferometer bearing discriminator and is presented to show typical circular interferometer techniques. This system makes use of four broadband antennas to cover 360 degrees in azimuth. To understand the operation, consider a signal from the north antenna that enters the matrix at $n = 1$. It passes through $\Sigma - H_1$ and $\Sigma - H_2$, appearing at $N = 0$ unchanged in phase (ignoring amplitude). This same signal is unchanged in phase at all other N ports. (By similar tracing, it may be seen that all four antenna inputs arrive at $N = 0$ with no phase change.) At $N = \pm 1$ the north Σ signal goes through

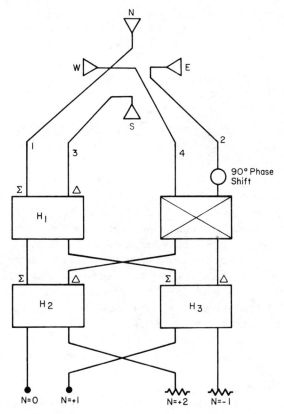

Figure 6-8. Four-element Butler-fed circular array.

a 90 degree shift with respect to $N = \pm 1$, to a 180 degree shift with respect to $N = \pm 2$ and a 270 degree shift with respect to $n = 1$. Table 6-1 shows the ultimate results for all signals.

By comparing the bearing angle of $N = 0$ to $N = \pm 1$ (the $N + 2$ and $N - 1$ ports are unused), the phase output will equal the phase input for a 1:1 correspondence. Any signal appearing in between the antennas will interpolate. The system

Table 6-1. *Phase Relationships at Output Ports N_n due to Antenna Inputs at n_n*

ϕ Input Direction	Phase Output at Port n				Δ
	$N = 0$	$N = +1$	$N = +2$	$N = -1$	$n = +1$ to 0
N (0 degrees) $n = 1$	0	0	0	0	0
E (90 degrees) $n = 2$	0	90	180	270	90
S (180 degrees) $n = 3$	0	180	0	180	180
W (270 degrees) $n = 4$	0	270	180	90	270

not used

is frequency independent and able to operate over a wider range; however, more elements are needed to improve the discrimination. This is in contrast to the narrowband, high-precision linear interferometer, which usually requires superheterodyne receivers with matched phase characteristics.

The system described above can be extended into a multielement format and, indeed, the Butler-fed circular array described in the previous chapter can be considered to be of that form, although grouped as an array antenna.

6.6 SUMMARY

Interferometer systems operate well over limited frequency ranges, providing the capability of receiving signals with greater accuracy despite the need for ambiguity resolution. As has been shown here, a combination of phase and amplitude monopulse techniques can be used to accomplish this latter objective. The popularity of the interferometer device derives from the very high accuracy the technique can achieve, limited only by the spacing and resolution problems. Some antenna structures, such as the spiral, can be tracked in phase over wide bandwidths, permitting interferometers to be constructed around host vehicles minimizing grounding effects. Most interferometers are of the superheterodyne configuration, offering high sensitivity with superheterodyne limited instantaneous bandwidths.

The simple configurations shown in this chapter may be extensively expanded to provide greater coverage and, as mentioned above, the Anaren circular array can be considered to be typical of most circular designs. The technique of moding is only one method to measure the angles. Direct comparison of the output of each element of the feed networks may be also used. In planar applications, interferometers may be mounted diagonally with respect to the horizon to provide both azimuth and elevation coverage. Different interconnections of the feed network can permit the polarization of an incoming signal to be measured. For these reasons, much of the future art of direction finding will undoubtedly make more use of interferometer techniques.

REFERENCES

1. Lipsky, S. E., "Find the Emitter Fast with Monopulse Methods," *Microwave*, May 1978.
2. Bullock, L., G. Oeh, and J. Sparagna, "An Analysis of Wide Band Microwave Monopulse Direction Finding Techniques," *IEEE Transactions on Aerospace and Electronic Systems*, Vol. AES-7, No. 1, Jan. 1971.
3. Hansen, R. C., *Microwave Scanning Antennas, Vol. III, Array Systems*, New York; Academic Press, Section V, 1966.
4. Doyle, W. C., "Electronic Warfare, Electromagnetic Support Measure and Direction Finding," *Signal*, Nov. 1971.
5. Baron, A., K. Davis, and C. Hoffman, "Passive Direction Finding and Signal Location," *Microwave Journal*, Sept. 1982, pp. 59 ff.
6. Baron, A., and C. Hoffman, "Navy E-2C Hawkeye Passive Detection System," *Journal of Electronic Defense*, Jul./Aug. 1981.

Chapter Seven

Methods for Signal Detection

In the discussion and examples of DF systems so far, noise and its effect on system false alarm rate and DF accuracy has been deliberately ignored. Here we will address noise and where it comes from in various receiver configurations.

Broadly speaking, noise can be considered a source of error whose effect is tolerable as a function of acceptable sensitivity, accuracy, and false alarm rate. If a DF system is to be capable of receiving signals with maximum sensitivity, noise becomes a limiting factor. A noisy signal or a signal that is below the noise level can be detected by exchanging time and/or bandwidth for sensitivity to provide detection. This solution is called processing gain, which is achieved by signal averaging (time), system adaptation (changing bandwidths) and by applying interactive mathematical methods, such as auto-correlation. The processing gain solution, however, takes time, during which intercept probability is reduced, not always an important factor in ELINT systems but essential for threat warning.

The loss of time during which a system is disconnected or unable to respond to signals is defined as shadow time. This time is important because nonrecurring individual pulse trains or short burst signals can be missed during this period. The time that is required to measure and present a new set of data either to an operator or to a decision-making system is defined as refresh time, the rate at which the operator of the receiver sees the display change or, in an automatic system, the time it takes to determine a significant change in the signal environment. If this time is too slow, there is the possibility that the system will be unable to perform its warning function. In contrast, an instantaneous, wide-open system response may allow random noise to trigger the threshold too often if it is set at too low a level. High threshold settings reduce these false alarms at the price of lower sensitivity by requiring higher signal-to-noise ratios.

A new problem occurs when high-density multiple-signal environments are encountered (which is practically always). Descriptors, used to recognize and separate individual signals from the environment, have to be measured quickly. Bearing and frequency are most commonly used since these can be measured accurately, permitting correlation or grouping of individual intercepts with the same characteristics. Most tactical threat warning receivers use mono- or single-pulse DF measurement since DF information cannot readily change on a pulse-by-pulse basis; it takes a finite time for an RF radiating aircraft missile or ship to change its course. Frequency, which can be measured to a higher accuracy, can, however, be rapidly changed, necessitating wide instantaneous RF receiver bandwidths or hyperfast tuning techniques (such as Microscan) to provide a reasonable probability of intercept. Under these conditions, computer processing systems may be unexpectedly overloaded.

Digital processing depends upon probable certainty in its binary decision-making process. Consider, for example, the case of a scanning radar being received on a pulse-by-pulse basis. As the pulses build from a low level with poor signal-to-noise ratio to a high level with good signal-to-noise ratio due to the scan, a low receiver threshold (defined as the level above noise above which a signal is deemed present), could falsely indicate many signals at many different directions since the variation in signal-to-noise ratio will change the probable accuracy of the direction measurement as a function of received signal strength. This is an acute problem for a DF driven receiver since the memory will erroneously fill storage spaces, or bins, with many bearings from the same signal, when in reality there is only one. An operator, by contrast, viewing a CRT display of actual DF video, would eliminate this problem by averaging the angle of arrival by mentally dividing the area occupied the many false strobes on the display, thereby providing better accuracy, effectively adding processing gain by the integration of many pulses.

7.1 CRYSTAL VIDEO RECEIVER

The most practical method to evaluate a systems signal-to-noise ratio starts with a very basic measurement: the tangential sensitivity or (TSS). For this measurement, a simple crystal video receiver, shown in Figure 7-1, is connected to a scope that displays the video output signal-plus-noise to noise ratio of the detected RF input signal. (The RF bandwidth B_r is usually much greater than the video bandwidth B_v for nonsuperheterodyne wide bandwidth receivers.) The RF source, which is modulated by a rectangular pulse capable of rising to full peak amplitude within the video bandwidth, is adjusted until the picture shown in Figure 7-2 results, which is the case where the bottom of the signal-pulse-noise level just touches the top of the

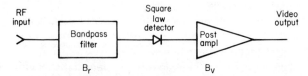

Figure 7-1. Simple crystal video receiver.

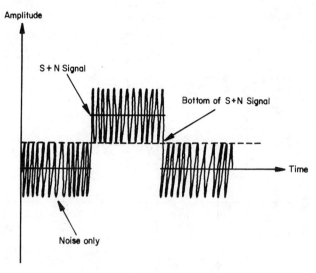

Figure 7-2. Tangential sensitivity display.

noise-only level representing the condition of an 8 dB video signal-to-noise ratio. The detector is by definition operating in its square-law region; therefore, the 8 dB observed video signal-to-noise ratio is actually a 4 dB RF signal-to-noise ratio (1). This measurement is defined as the tangential sensitivity. Values of RF input of -50 dBm for video bandwidths of 2 MHz are typical.

The simple case described above is a gain-limited crystal receiver since there is no predetection RF gain, and varying the RF bandwidth from B_r to infinity will not change the signal-to-noise ratio output, which is determined by the noise of the crystal detector and video amplifier. Varying the bandwidth of the video amplifier, however, will reduce the noise by $1/B_v$ and the sensitivity by $1/\sqrt{B_v}$ as the result of the square-law detector characteristics. Doubling the video bandwidth will cause a 3 dB output increase in noise. A 1.5 dB increase in the RF predetection input level is necessary to produce a corresponding 3 dB output video signal increase.

Unfortunately, measurements for signal-to-noise ratios based upon tangential visual measurements are very subjective, and it is possible to obtain widely divergent results from different observers. Additionally, the TSS 8 dB video signal-to-noise ratio is not a useful threshold for most systems since the false-alarm-rate at this level is too high. The value in the use of TSS lies in its relative ease of measurement and the utility of fixing one point in the dynamic range of a system from which to calculate various noise effects.

To obtain higher sensitivities, an RF amplifier of sufficient gain and a noise figure less than that of the detector/video amplifier can be placed ahead of the simple detector to improve receiver sensitivity (2). The same effect is obtained in super-heterodyne receivers by using the gain of the IF amplifier in conjunction with the improved conversion efficiency of a mixer. There are significant differences in the

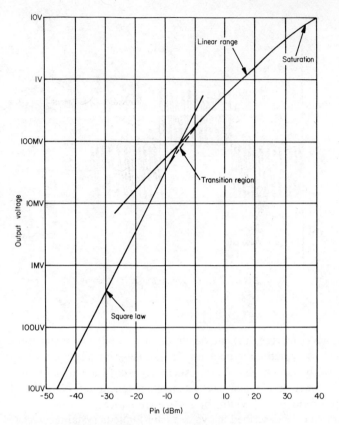

Figure 7-3. Detector diode—typical characteristics.

two approaches; however: In the RF amplified crystal video system, the detector is operated in its square-law region; in the superheterodyne, the detector is operated in its linear region. Figure 7-3 shows the RF-to-video transfer curve of a typical crystal detector (3). It may be seen that the square-law range extends from −45 dBm to approximately −7 dBm, achieving an RF dynamic range of about 35 dB. The detector becomes linear at approximately −10 dBm to about +40 dBm. Two other important differences exist: In the RF preamplified crystal video receiver, the RF bandwidth B_r is much greater than the video bandwidth B_v; in the superheterodyne, the B_r may be only equal to or only slightly greater than twice the video bandwidth ($B_r \cong 2B_v$). These features are desirable for passive direction finding on their own merits: The wide RF bandwidth gives maximum probability of detection; the narrow RF bandwidth superheterodyne helps in dense signal environments. All of these considerations make determination of the expected sensitivity of a receiver difficult to predict unless great care is taken to consider all effects properly.

7.1.1 Video Detectors

The gain-limited crystal video receiver is defined as such by the fact that the output signal-to-noise ratio is due primarily to the noise generated in the crystal detector and video amplifier input resistance, making the sensitivity dependent on the bandwidth of the video amplifier. This is the condition of insufficient gain ahead of the detector to establish the output signal-to-noise ratio as a function of a predetection element such as an RF amplifier.

Gain-limited crystal receivers offer the feature of simplicity at the price of low sensitivity since the equivalent "noise figure" of the simple receiver is poor due to inefficiency of the crystal and problems of matching it to the RF circuit and the video amplifier.

Figure 7-4 is a diagrammatic representation of a diode detector in an RF mount, which has an RF input impedance of typically 50 ohms (4). C_j and R_j are the diode junction capacitance and resistance, R_s and L_s are the ohmic base spreading resistance and lead inductance; C_p is the parallel capacitance across the holder. The dynamic video resistance of the diode is

$$R_v = R_s + R_j$$

The voltage current equation of the diode is

$$i = I_s \exp\left[\left(\frac{q}{nkt} V_j\right) - 1\right] \qquad (7\text{-}1)$$

where

$$i \quad = \quad \text{output current}$$

$$I_s \quad = \quad \text{the diode saturation current}$$

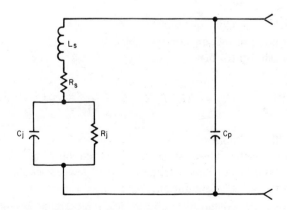

Figure 7-4. Diode detector equivalent circuit.

V_j = the voltage across the diode junction

K = Boltzmann's constant

     = 1.38×10^{-16} erg/K

T = the absolute temperature in degrees Kelvin

q = the electrical charge constant (1.6×10^{-19} C)

The number n is referred to as the "nonideality" factor (5) and may vary from 1 to 1.08 for Schottky detectors. The dynamic video resistance of the diode may be obtained by differentiating (7-1) and solving for $R_j(i)$, which yields

$$R_j(i) = \frac{nkt}{q(I_D + I_s)} \tag{7-2}$$

where

I_D = a forward bias current usually applied to the diode to improve its RF impedance match

I_s = the saturation current of the diode ($\approx 10^{-9}$ A)

The value of R_j varies with signal level, bias, and temperature. The purpose of the above calculation is to obtain an equivalent resistance for the specification of the total noise in the receiver.

The sources of noise in the simple video detector consist of the following components:

1. Thermal noise of R_s in bandwidth B_v

$$N_T = 4KTB_vR_s$$

where $B_v \cong .35$/pulse rise time

2. The shot noise of the junction resistance R_j
It may be shown (6) that

$$N_s = 2KTB_vR_j\left(1 + \frac{I_s}{I_D + I_s}\right)$$

3. The $1/f$ "flicker" noise
This low-frequency noise is generally ignored since most video amplifiers are bandpass devices with low-frequency cutoff frequencies above the frequency at which this effect is significant.

4. The noise power contained in a resistance representing the input impedance R_A of the video amplifier

$$N_A = 4KTB_vR_A$$

Ignoring the flicker noise, the total output noise voltage is

$$V_N^2 = 2KTB_v \left[2R_s + R_j \left(1 + \frac{I_s}{I_D + I_s} \right) + 2R_A \right] \qquad (7\text{-}3)$$

If the noise contributed by the amplifier can be considered negligible when compared to the diode, then

$$V_N = \sqrt{2KTB_v \, 2 \left[R_s + R_j \left(1 + \frac{I_s}{I_D + I_s} \right) \right]} \qquad (7\text{-}4)$$

There have been many attempts to define a figure of merit M for a detector, equivalent to noise figure. Reference 7, an important paper on tunnel diode detectors, defines it as follows:

$$M = \frac{K_D}{\sqrt{R_v}} \qquad (7\text{-}5)$$

where

R_v = the dynamic resistance of the diode (per above)

K_D = the open-circuit voltage sensitivity of the device in millivolts per milliwatt (mV/mW)

For one type of detector, the tunnel diode K_D can range up to values as high as 3000, with R_V values at about 300 ohms. This type of detector operates with a low conversion loss at low signal levels and a high conversion loss at high signal levels. (If a matching attenuator pad is inserted in series with the detector to effect a better VSWR match, it will add twice its value to all conversion losses, thus disproportionately reducing performance.)

On the other hand, Lucas, in a classic paper (8), defines a figure of merit,

$$M_s = B_s \sqrt{R} \qquad (7\text{-}6)$$

where

B_s = the short circuit current sensitivity of the detector

R = a load resistance equal to the video resistance

This case assumes no noise contribution from the video amplifier, which is often reasonable. Using this value of M_s, the tangential sensitivity of a simple crystal receiver with no preamplification is

$$t_s = \frac{1}{M_s} \sqrt{4KTB_v \ (6.31) \ (f + t - 1)}$$

$$= \sqrt{\frac{4KTB_v}{M_s^2} \ 6.31 f_v} \qquad\qquad (7\text{-}7)$$

where

t_s = the input RF signal power required to achieve TSS

B_v = the bandwidth of the video amplifier

K = Boltzmann's constant

t = noise temperature ratio of the detector

T = temperature in degrees Kelvin

f = noise figure of the video amplifier

f_v = effective noise figure of the video amplifier and detector

Solution of this equation yields typical TSS values in the order of -48 dBm using a low-noise video amplifier with a 2 MHz B_v.

From these equations it may be seen that all of the noise of the gain-limited receiver is produced by the detector and video amplifier and that its effective noise figure is quite high. The signal level necessary to overcome this noise will need to be fairly substantial, placing it well above the noise in the antenna. Any gain in the antenna, however, will improve the system's overall sensitivity compared to an isotropic radiator by acting as an ideal amplifier. Maximum sensitivity will be achieved when the RF voltage across the detector junction is at a maximum. This occurs when the detector is well matched to the source (50 ohms). To achieve this match, diode detectors may be biased, or the detector may be transformed by a matching network to the 50 ohm RF line impedance. If operated with zero bias, a higher VSWR between the detector and the signal source will result, since the diode dynamic impedance increases reducing power transfer. The relationships are very complicated, and although attempts have been made to define a figure of merit for a simple diode, the proliferation of diode types has not resulted in agreement of a single definition for video receivers.

Point-Contact Silicon Diodes. Point-contact silicon detectors, dating from radar technology, were the first available types. These detectors, which use a metallic "cat whisker" to contact a semiconductor, exhibit video dynamic resistances of approximately 20,000 ohms, dropping to 2000 ohms with approximately 200 μA of forward bias, which also improves the RF match (lower VSWR) (9). In the days of vacuum tubes, these resistances permitted high-input impedance amplifiers to act as potential measuring voltmeters, effectively allowing a maximum voltage transfer to be obtained between the detector and the amplifier's relatively high-input impedance. The limiting factor was the maximum video bandwidth which was limited due to upper frequency rolloff from distributed and parasitic capacity developing a reactance equaling the tube's high-input impedance thus making B_v too low for optimum pulse fidelity. The voltage potentiometer technique, however, allowed B_v to be narrowed for extremely good sensitivities, permitting TSS values of -70 dBm to be obtained for very narrow basebands starting at DC, and useful for CW detection. Transistor circuits, by contrast, are essentially low-impedance devices, and the maximum voltage match obtained above cannot be easily obtained. Additionally, point-contact diode impedance becomes lower as the RF power increases. On the positive side, the point-contact diodes have good high-frequency response due to extremely low junction capacitance. With modern semiconductor technologies, however, this is no longer significant. The flicker noise problem and the manufacturing unpredictability of the point-contact detector has reduced its popularity.

Tunnel Diodes. Tunnel or back-diode detectors biased near their peak forward current provide the highest rise time response and good sensitivity (expressed in millivolts per milliwatts). Values of 3500 are typical for low-level signals (7). Although tunnel diode detectors start to depart from square-law at about -20 dBm, their improved sensitivity tend to compensate for this by providing approximately the same dynamic range. Tunnel diode detectors operate with low-video resistance loads (≈ 300 ohms) and at zero bias, since bias increases the video resistance as compared to point contact detectors. This operating resistance is fairly constant over the dynamic range. The tunnel diode low-video impedance assures wide bandwidths for pulse fidelity, accounting for their improved performance in wide B_v applications. They are about equal to other detectors in narrow video bandwidth applications; however, the unbiased tunnel diode offers a $1/f$ corner frequency considerably (about 3 decades) below that of the crystal detector permitting them to be used for DC detection of CW signals, doppler radar, and zero frequency IF applications. Tunnel diode detectors have another unusual property: When matched at low levels they are able to withstand high power well since the reflectivity of the diode becomes greater with increasing RF input.

The Schottky Diode. The Schottky barrier or hot-carrier diode is another detector that is becoming popular due to its fabrication consistency, very high sensitivity, and low-impedance characteristics. This type of device develops its diode action by the flow of electrons from a semiconductor (n type) or by the flow of holes (p type) to a metallic deposited conductor. As shown in Figure 7-5, a "depletion layer" is formed in the manufacture of the diode as electrons or holes flow from the

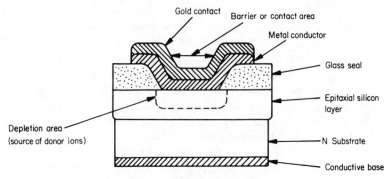

Figure 7-5. Planar Schottky diode.

semiconductor to the metal. This layer (for n material) contains ions that, free of their free electrons, act as positive "donor" atoms, creating a positive potential between the N substrate and the metallic conductor. The donor action continues until the depletion layer is formed, resulting in about 0.5 V potential difference when connected so that when a positive voltage is applied to the metal, the internal voltage falls and electrons flow into the metal creating the diode action. The electrons that have greater than average thermal energy escape into the metal and are called hot carriers. Since no electrons (minority carriers) flow from the metal to the semiconductor, the current stops almost instantly, establishing a reverse voltage in a subnanosecond period. This action eliminates stored charge problems found in other diodes, permitting Schottky diodes to operate at millimeter wave frequencies. Reference 10 gives an excellent description of this concept. The Schottky diode operates with bias and can exhibit greater sensitivity; however, its rise time is not as fast as tunnel diodes. The width of the barrier or contact surface is more predictable than a point-contact diode. It is interesting to note that a point-contact diode, which is really a metal-to-semiconductor junction, operates in the same manner as the Schottky although it does not follow the theory as well due to fabrication variances (11).

7.1.2 Square-Law Crystal Video Receiver Sensitivity

The wide instantaneous RF bandwidth of the crystal video receiver assures high-intercept probability. We have seen, however, that signals must be of sufficient level to overcome the inherent noise of the detector/video amplifier. Our definitions have related to the concept of tangential sensitivity as a convenient output level from which to draw comparisons. As was stated previously, tangential sensitivity representing an 8 dB signal-to-noise ratio is not suitable for threshold systems since noise spikes will cross an 8 dB threshold very often (several times per second) causing serious false alarms. What is required is the addition of RF gain ahead of the simple detector to establish a signal-to-noise ratio that is less noise dependent on the detection-video amplification process. Unfortunately, this cannot be done simply for the video signal-to-noise ratios desired due to the square-law action of the detector.

Ayer (2) approached the problem of adding gain to a detector by using a broadband low-noise traveling wave tube (TWT) amplifier. Starting with the minimum discernible signal and with no preamplifier gain, he found that sensitivity increased on a decibel-by-decibel basis as gain was added by the TWT until a maximum sensitivity point was reached past which there was no further improvement. From this he developed the concept of an effective bandwidth, B_e. He derived an equation for this bandwidth relating the RF bandwidth B_r and video bandwidth B_v by integrating the noise spectrum across the RF bandwidth B_r in one cycle line spectrum intervals. By performing this integration after square-law detection and by eliminating terms, the equation for an equivalent bandwidth B_e results:

$$B_e = \sqrt{2B_v B_r - B_v} \qquad (7\text{-}8)$$

This equation, which was derived for the minimum discernible signal $(S + N/N)_o = 1$, considered some of the effects of amplified noise but did not fully cover the many different signal and noise cross-products conditions encountered over the dynamic range from TSS to gain independence for various ratios of B_v to B_r. Unfortunately, it is not possible to use this concept of an effective bandwidth, except for a very limited case, due to the complicated small signal suppression effects resulting from complex noise and signal cross products occurring in the RF bandwidth; these effects varying as a function of the signal levels and the ratios of B_r to B_v. It is important to know how to compute the sensitivity of a RF preamplified crystal video receiver, or Ayer's equation may be misapplied, resulting in serious errors.

When sufficient RF gain is applied ahead of a crystal receiver, as shown in Figure 7-6, the receiver will become noise-limited. This is the condition where the predetection gain is sufficient to ensure that the noise output, in the absence of a signal, is due to detection of the noise figure multiplied by its gain referenced from the *KTB* source noise present in the amplifier's input and source impedance. Since the noise figure of microwave RF amplifiers is invariably lower than the equivalent noise figure of the simple crystal receiver, a dramatic sensitivity improvement is to be expected, and indeed the noise-limited receiver is most desired for passive DF applications for this reason. Figure 7-7 is a typical plot of the improvement to be expected by the addition of RF preamplification to a crystal video receiver. A

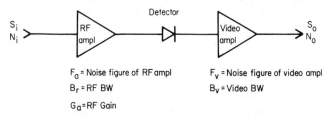

F_a = Noise figure of RF ampl F_v = Noise figure of video ampl

B_r = RF BW B_v = Video BW

G_a = RF Gain

Figure 7-6. Amplified crystal video receiver.

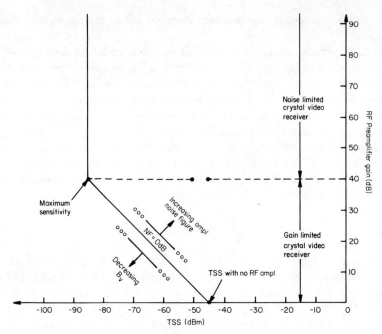

Figure 7-7. Effect of adding gain to a simple crystal video receiver.

typical TSS of -45 dBm is shown for the case of no predetection gain. As gain is added, there is a corresponding increase in sensitivity until the receiver is noise limited, at which point (shown as 45 dB gain case here) no additional gain will effect a sensitivity improvement. [The amplifier is shown idealized (noise figure = 1); practical noise figures will reduce the sensitivity as shown.] Correspondingly, decreasing video bandwidths B_v (for a fixed RF bandwidth B_r) will increase the sensitivity. It is important to note that the addition of gain decreases the dynamic range of the system at the expense of using up the square-law region of the detector. There is seldom any virtue in utilizing any more gain than is necessary since it is also expensive. The transition from the gain-limited receiver to the noise-limited receiver is approximately the point after which Ayer's Equation (7-8) can be used (when $B_r > B_v$), as will be shown. There are, however, several problems with this approach over and above the obvious requirements of gain and/or phase matching for DF measurement. The action of the square-law detector plays an important part in the determination of the output signal-to-noise ratio of the receiver for a given input signal level. The output of a detector for the conditions of noise alone and for a signal-plus-noise are depicted in Figure 7-8. Assuming that the input noise is Gaussian and the gain G_A of the RF amplifier is sufficient to attain the noise-limited condition, the video output noise-power spectrum will be of a triangular shape as shown in the shaded section of the figure. This spectrum results from the sum of all of the noise $\times$ noise products of the RF amplifier with a bandwidth B_v occupied within the bandwidth B_r. The noise output and the integrated power of all of these products are superimposed within the video bandwidth B_v of the video amplifier due to the RF noise interaction in the nonlinear detector.

If a signal is added to the noise, the area occupied by the sum of the signal and noise power $s_i \times n_i$ becomes all of the area under the upper curve, which is a considerably greater total than the case of noise alone above. In the $s_i \times n_i$ case, it can be seen that the sum of the two triangular areas equals the area of the n_i case above; therefore, the square area is the effect of the presence of the signal $\times$ noise products and the signal. For small input signal-to-noise ratios, the output signal-to-noise ratio varies as the square of the input. For large input signal-to-noise ratios, the output varies directly. This phenomenon is identified in Davenport and Root (1) as the "small signal suppression effect." For large signals, the output is due mainly to interaction of $s \times n$, whereas for small signals, the output of $n \times n$ becomes a more significant portion of the detected output, suppressing the small signal. This phenomenon was not considered by Ayer even though he was working with small signal ($s + n = n$) output case. Since in most applications output signal-to-noise ratios greater than TSS (8 dB) are required, the corresponding output signal-to-noise ratios are more exactly related to the input. Indiscriminate use of Equation (7-8) will give pessimistic results for these large signal-to-noise ratio cases.

Many papers have been written attempting to establish the methodology of determining the sensitivity of amplified crystal video (and superheterodyne) receivers. Rice (12), Lucas (8), and Klipper (13) have developed mathematical relationships that are based initially upon certain subjective criteria such as TSS or minimum discernible signal level as the basis of formula deviation. In many cases, these equations tend to yield results that are inconsistent with measured values of a completed system.

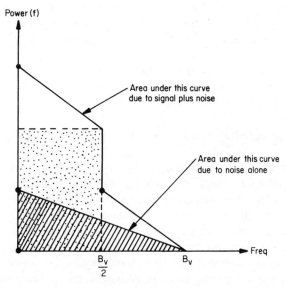

Figure 7-8. Video output power spectrum of the RF amplified crystal video receiver.

The most comprehensive paper on sensitivity, written by Lucas, offers equations based upon tangential sensitivity assumptions (RF signal-to-noise = 4 dB) for noise performance. Unfortunately, the results have been difficult to use. Tsui (16) has developed a more flexible formulation for the TSS case of a crystal video receiver for both the gain and noise limited crystal receiver based upon Lucas.

Note: In the work that follows, all sensitivities will use decibels [10 log (ratio)] instead of the direct ratio formats. Noise figures and system losses will therefore be negative. Gains and bandwidths will be positive, and all sensitivities will assume variation from the noise present in a 1 MHz bandwidth; that is,

$$\text{the noise in 1 MHz} = 10 \log \text{KTB} = -114 \text{ dBm/MHz}$$

where

$$B = 1.0 \text{ MHz}$$

$$T = 290 \text{ K}$$

$$K = \text{Boltzmann's constant}$$

An increase in bandwidth (BW) will be computed as a positive value

$$10 \log(\text{BW})$$

and added. All lowercase values, for example, f, will refer to ratios in times, whereas uppercase values will refer to this ratio in decibels; for example, $F = 10 \log f$ for predetection and $F = 20 \log f$ for postdetection due to squaring by the detector. The logarithm of a square root is

$$10 \log \sqrt{x} = 5 \log x$$

and the log of a product is

$$\log(A)(B + C) = \log A + \log(B + C), \text{ etc.}$$

The tangential sensitivity Equation (7-7) can be written in a more general form by considering the following: The equation expressed the tangential sensitivity of the simple crystal receiver in terms of the effective noise figure of the crystal detector and video amplifier. If an RF amplifier is added, it provides a gain (g_r) and a noise figure (f_r), the gain being sufficient to allow f_r to appear on the detector's transfer curve at a point greater than the tangential sensitivity of the simple receiver. This will add the noise contained in the RF amplifier bandwidth, which when detected, will add $n \times n$ and $s \times n$ noise in the video bandwidth B_v. This must be accounted for by adding Ayer's Equation (7-8) to Equation (7-7). Ayer, in the computation of the effective bandwidth B_e, considered only the case of the signal power being equal to the noise, requiring the addition of another term (2.5 times) to account for the predetection value of the tangential sensitivity as defined.

The total tangential sensitivity equation, expressed in decibels, has been restated by Tsui (14) as follows:

$$T_s = -114 + F_A + 10 \log \left(6.31 B_v + 2.5 \sqrt{B_e^2 + \frac{AB_v}{(g_A f_A)^2}} \right) \quad (7\text{-}9)$$

where

$$B_e = \sqrt{2B_v B_r - B_v^2} \quad \text{(from Eq. 7-8)}$$

F_A = the RF amplifier noise figure in decibels (dB)

g_A = the RF amplifier gain

6.31 = the ratio for TSS (10 log 6.31 = 8 dB)

B_v = the video amplifier bandwidth

B_r = the RF bandwidth (usually defined by the preamplifier)

and

$$A = \frac{4 f_v R_v}{KTM^2} \times 10^{-6} \quad (7\text{-}10)$$

where

f_v = the noise figure of the video amplifier

R_v = dynamic detector video resistance

M = the detector figure of merit, defined in Equation (7-5)

This general equation allows determination of T_s for both the noise and gain-limited case. The most notable effect when adding an RF preamplifier is related to the terms under the square root sign. Considering the most useful case of $B_r \gg B_v$, then the effective bandwidth term becomes more significant than the second term and for the gain-limited case (defining T_{sg} as the gain-limited T_s):

$$T_{sg} = -114 + F_A + 10 \log(6.31 B_v + 2.5 B_e) \quad (7\text{-}11)$$

$B_e \gg B_v$ for most cases since

$$B_A \gg 2B_v$$

$$= -114 + F_A + 10 \log 2.5 B_e$$

$$T_{sg} = -110 + F_A + 10 \log B_e \qquad (7\text{-}12)$$

This is a most useful form and can be used to readily determine the gain-limited tangential sensitivity. The minimum or confining gain defined as the gain G_A necessary to just assure gain-limitation occurs when

$$B_e = \frac{AB_r}{(g_a f_a)^2}$$

Solving for this gain in terms of T_{sv} gives

$$G_A = 110 + T_{sv} - F_r - 10 \log B_e \qquad (7\text{-}13)$$

Solving Equations (7-12) and (7-13) simultaneously to obtain the gain-limited tangential sensitivity T_{sg} yields

$$T_{sg} = T_{sv} - G_A \qquad (7\text{-}14)$$

In other words, *the tangential sensitivity of the simple crystal video receiver is directly improved decibel-for-decibel as long as at least the minimum confining gain G_A is used.* This is not exact, however, since there is still the small signal suppression effect to consider.

Let us now consider the noise-limited case. This is readily done by rewriting Equation (7-9) with $F_A = 0$ dB ($f_A = 1$) and $G_A = 1$.

Then

$$T_{sr} = -114 + 10 \log(6.31 B_v + 2.5 \sqrt{AB_v}) \qquad (7\text{-}15)$$

or approximately

$$T_{sr} = -110 + 10 \log \sqrt{AB_v} \qquad (7\text{-}16)$$

In this case, A must be determined by using the diode parameters from a vendor's data sheet or by actual measurement.

The above equations permit the determination of the tangential sensitivity of both noise-limited and gain-limited crystal video receivers. The equations do not take into account losses that may precede the receiver such as are found in preselectors, filters, cables, or the potential system isotropic antenna sensitivity gain when using high-gain horns or other types of antennas. Since most crystal video applications use spiral antennas, with nominal 0 dBi gain, it is easy to forget the improvement a high-gain steerable reflector antenna or multichannel array can provide.

The use of tangential sensitivity derives from World War II methods and techniques and is popular because it provides a historical method of comparison of the technological advancement in detectors as well as a means of comparing contemporary designs. More automated production testing methods rely on the presence of detector DC shift for a CW signal for accurate sensitivity characterization and parametic measurement of intrinsic capacitance resistance and inductance for high-frequency performance. In most receiver applications, output signal-to-noise ratio sets the system performance. Values of about 15 dB or more are usually required for proper threshold crossing detection with typical false alarms of about 1 per minute. For these reasons, the use of the figure-of-merit type comparisons at tangential sensitivity alone is becoming limited, although it is convenient to measure.

The next section presents a signal-to-noise technique that permits sensitivity determination for a wide range of output signal-to-noise ratios, and since the tangential signal-to-noise ratio is a convenient defined point, it will be used to compare results. In practical situations, tangential sensitivity should be used in conjunction with other performance factors such as detector conversion efficiency, dynamic range, RF impedance, match, bandwidth, and operating input and output impedances to specify total system performance. The equations and methods presented here facilitate this.

7.1.3 Signal-to-Noise Sensitivity Methods

The preamplified crystal video receiver is relatively easily analyzed for the tangential output signal-to-noise ratio. It is far more useful to develop a general signal-to-noise output-to-input transfer characteristic to account for a wide range of conditions. Unfortunately, there has been much disagreement about the methods of this calculation resulting from various interpretations of the works of Rice (12), Lucas (8), and Klipper (13), A first published attempt at clarification, published by this author (15), has resulted in much correspondence and evaluation, the results of which are presented here.

For a gain-limited amplified crystal video receiver shown in Figure 7-6, the noise spectrum is of the type that was shown in Figure 7-8. There is an outside envelope that reaches a maximum at DC and tapers off in a triangular fashion reaching a zero at the limits of B_v. This describes the base-band noise output, assuming that the video amplifier is essentially bandpass filtered. Depending upon the low-frequency cutoff effects, the actual noise output is the result of adding the two noise × noise triangular areas and the one rectangular area shown in the figure. In the absence of a signal, there is a constant noise triangular area extending from DC to B_v. The calculated value of the area bounded by the outer limit to the area defined by the noise above the triangle gives the output video signal-to-noise ratio. Since the detector is operating in its square-law region, the various areas are highly dependent upon the amplifier noise figure F and input RF signal-to-noise ratio. For small signal-to-noise input ratios (<1), the $n \times n$ products dominate the output, whereas for larger signal-to-noise ratios the $s \times n$ product is dominant. This relationship is therefore dependent upon the law of the detector, the signal-to-noise ratio, and the ratio of the video to RF bandwidth. Because of these dependencies, general rules are hard to state, and for this reason amplified crystal video receivers have earned a reputation for unpredictable sensitivity calculations. The methods followed here will attempt to show how more exact computations can be made; however, much of the theory depends upon assumptions that are not always exact. (An example of this is the assumption that all of the noise in the RF bandwidths is Gaussian and that the shape of the RF bandpass is the same.) Derivation from the tangential sensitivity case is also somewhat subjective.

Calculation of the noise and signal cross-product areas for the half-wave detector has been done by each of the previously cited references and lead in general to an equation that shows the video output signal-to-noise (S/N) ratio X_o as a function of the RF input signal-to-noise ratio X_I of the form

$$X_o = \frac{X_I^2}{a + bX_I} \tag{7-17}$$

where a and b are constants defined as follows:

$$a = \begin{cases} 2\dfrac{B_v}{B_r} - \left(\dfrac{B_v}{B_r}\right)^2 & \text{for } B_v < B_r \\ 1 & \text{for } B_v > B_r \end{cases}$$

$$b = \begin{cases} 4B & \text{for } B_v < B_r/2 \\ 2 & \text{for } B_v > B_r/2 \end{cases}$$

The most useful form of the general equation is for $B_r >> B_v$ or

$$(S/N)_o = \frac{(S/N)_I^2}{2\dfrac{B_v}{B_r} - \left(\dfrac{B_v}{B_r}\right)^2 + 4\dfrac{B_v}{B_r}(S/N)_I} \tag{7-18}$$

This equation represents the usual case of an RF bandwidth as defined by a lossless filter greater than the video bandwidth and has been plotted in Figure 7-9. The video output signal-to-noise ratio is given as a function of the RF input signal-to-noise ratio range for various ratios of B_v/B_r.

General Method. To determine the sensitivity of a noise-limited receiver, the best procedure is to start with the desired output signal-to-noise ratio, determine the values of B_r and B_v, and graphically read the RF input signal-to-noise. To this value add all other losses (such as the loss in the filter that determines B_r) and the effect of gains due to amplifiers to arrive at the input RF signal-to-noise ratio. This

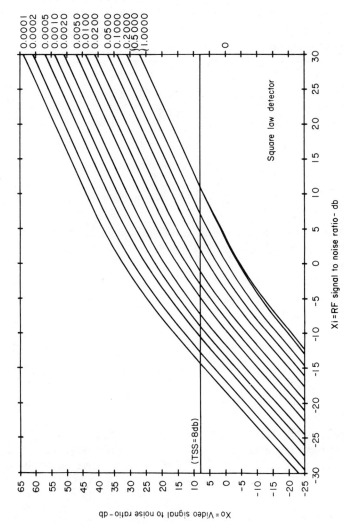

Figure 7-9. RF signal-to-noise ratio curves for various B_v/B_r values for a square-law detector.

189

procedure is most appropriate since the output signal-to-noise ratio is usually chosen to satisfy a required false alarm rate. The following example illustrates the general method using TSS as the desired output signal-to-noise ratio for the purposes of comparison. The curves, however, are designed to be used for the full range of output signal-to-noise ratios.

The 8 dB tangential sensitivity line has been drawn as follows:

$$\left.\begin{array}{l} B_v = 0.5 \text{ MHz} \\ B_r = 120 \text{ MHz} \end{array}\right\} \quad B_v/B_r = \frac{0.5}{120} = .004$$

$$F_A = 6 \text{ dB}$$

$$G_A = 50 \text{ dB}$$

$$T_{sr} = -44 \text{ dBm (with no preamplification)}$$

The desired output tangential sensitivity is 8 dB, which is entered on the abscissa and brought to the B_v/B_r curve of .004 where an RF input signal-to-noise ratio of -5 dB is read.

The input signal level to produce an 8 dB signal-to-noise video output level will be measured below the noise in the RF bandwidth B_r. The signal for an 8 dB (signal-to-noise) video output is

$$(S_I) = (\text{available noise in 1 MHz}) + F_A + (S/N)_I + 10 \log B_r$$

$$= -114 + 6 - 5 + 10 \log 120$$

$$S_I = 92.2 \text{ dBm}$$

Let us compare this value to that for the TSS video output case using the equations previously derived, recognizing again that we have only used $(S/N)_o = 8$ dB for this comparison; the method is general for all signal-to-noise values shown on the curve.

The TSS Method. To proceed, it is necessary to determine if the B_e value can be used directly or if the $6.31B_v$ term in Equation (7-11) must be considered; that is, $6.31B_v << 2.5B_e$ for B_e:

$$B_e = \sqrt{2B_vB_r - B_v^2} = \sqrt{(2)\,(.5)\,(120)}$$

$$B_e = 10.95 \times 10^6 \text{ and } 2.5B_e = 27.3 \times 10^6$$

Testing this against $6.31B_v = 3.15 \times 10^6$ shows that Equation (7-11) is preferable to Equation (7-12) since 27.3 is not $>> 3.15$.

Substituting in Equation (7-11)

Table 7-1. Tabulation of TSS Sensitivities for a Noise-Limited Crystal Video Receiver Using Three Methods

B_r (MH$_3$)	B_v (MH$_3$)	Initial TSS (dBm)	Amplifier (db) Gain	Amplifier (db) FA	Hughes Calculated TSS (dBm)	Graphically Determined TSS (dBm)	Measured TSS (dBm)
40	7.7	−41	51	5	−88.7	−86.5	−88
120	7.7	−41	48	6	−87.1	84.7	−86
40	0.5	−44	53	5	−95.0	−95.5	−95
120	0.5	−44	50	6	−93.0	−92.2	−92

$$S_i = -114 + 6 + 10 \log 30.5$$

$$S_i = 93.2 \text{ dBm}$$

comparing well with 92.2 above.

The gain of 50 dB when added to the T_{sr} of −44 dBm yields −94 dBm, which is also in reasonable agreement.

The results are also in accord with the value of −92 dBm measured and −93 dBm calculated for these conditions (see Hughes, 16).

It is interesting to compare other values as calculated by use of the curves to those calculated and measured by Hughes for various ratios of video and RF bandwidths. Table 7-1 tabulates these results. It may be seen that there is reasonable correspondence for all methods, with better correspondence where $B_r >> B_v$. In one case, $B_r = 120$, $B_v = .5$, there was more deviation; however, the amplifier gain was several decibels less.

The value of the general case is the wide range of values over which the computation can be made. The value of the tangential calculation method is the ability to predict a point that can be readily measured. It must be remembered that the sufficient confining gain case is always assumed to be true in the graphical approach.

7.2 SUPERHETERODYNE RECEIVER

The superheterodyne receiver has been in use since the dawn of radio communications and certainly before the development of radar. For the particular use in passive detection DF systems, we are generally interested in two basic forms: the preselected or "tuned" receiver and the broadband or "downconverted" configurations. This is not to exclude other designs such as the instantaneous frequency measurement, homodyne, or microscan configurations, each of which serves a purpose in the overall determination of signal presence and identity; however, the specific characteristics of the superheterodyne as used for DF purposes merit particular attention.

7.2.1 Sensitivity Determination Methods

Early World War II receivers utilized mechanically scanned RF preselectors tracked to a local oscillator, which could be either a tube or klystron type, to provide a "window" in the RF spectrum through which signals could enter and be analyzed. RF amplification was not available; therefore, the noise figure of the receiver was solely dependent upon the preselector loss and the conversion loss of the mixer, which, in turn, depended upon the noise figure of the first intermediate frequency amplifier. The concept of noise figure (F) was born as a ratio of the noise present in the actual receiver to that in a theoretically noiseless receiver at a specific temperature $T(\text{K})$:

$$F = \frac{\text{actual noise}}{\text{noise from ideal receiver at } T}$$

$$F = \frac{N_o}{GN_i} \tag{7-19}$$

where

$$N_o = \text{actual output noise}$$

$$N_i = \text{KTB noise in the receiver bandwidth } B$$

$$G = \text{receiver net gain}$$

Modern superheterodyne receivers may contain many active gain stages and may have loss elements such as preselectors in series with the mixing and detection process. Consider Figure 7-10, which shows a typical configuration. The overall noise figure f_t is

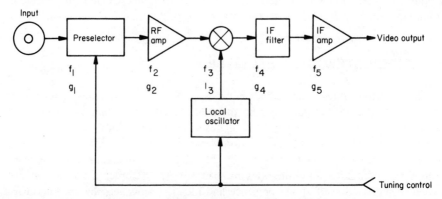

Figure 7-10. Basic narrow-band superheterodyne.

$$f_t = (f_1 \underbrace{\quad +}_{\boxed{\text{Preselector}}} \quad \underbrace{\frac{(f_2 - 1)}{g_1}}_{\boxed{\text{RF Amp}}} \quad + \quad \underbrace{\frac{(f_3 - 1)}{g_1 g_2}}_{\boxed{\text{Mixer}}} \quad + \quad \underbrace{\frac{(f_4 - 1)}{g_1 g_2 g_3}}_{\boxed{\text{IF Amp}}} \quad + \quad \cdots$$

where each f and g represents the noise and gain ratios.

Applying these gain and noise ratios to the computation of overall noise figure, the preselector exhibits a loss; therefore $f_1 = l_1$. The RF amplifier contribution is $l_1(f_2 - 1)$ since $g = 1/l$ for all values of $g < 1$. The mixer stage must include the effects of the IF amplifier or

$$\frac{(f_3 - 1)}{g_1 g_2} = \frac{l_1(f_3 - 1)}{g_2}$$

f_3 is the mixer loss and may be computed from

$$f_3 = l_c (l_4 f_5 + t - 1)$$

where

l_c = the mixer conversion loss

l_4 = the IF roofing filter loss

f_5 = the noise figure of the IF amp

t = nonideality factor of the diode (noise temperature ratio)

The equation for the total noise figure of the receiver can be written as follows:

$$f_t = l_1 + l_1(f_2 - 1) + l_1 \frac{[l_c (l_2 f_1 + t - 1) - 1]}{g_2} \tag{7-20}$$

assuming that terms past the first IF amplifier can be ignored. From these relationships, the overall noise figure can be calculated.

The concept of confining gain for the superheterodyne relates to supplying sufficient gain in one stage to reduce the noise contribution effects of the next.

As an example, from the equation associated with Figure 7-10

$$f_t = f_1 + \frac{f_2 - 1}{g_1}$$

to make the second term negligible so that $f_t = f_1$ or that the noise figure of the superheterodyne is that of the first stage

$$\frac{f_2 - 1}{g_1} \ll f_1$$

or g_c which is defined as the confining gain which attains our objective.

$$g_c \gg \frac{f_2 - f_1}{f_1} \text{ or approximately}$$

$$(7\text{-}21)$$

$$g_c \gg \frac{10(f_2 - f_1)}{f_1} \text{ as a good rule}$$

Some logical interpretations of the above equations can be drawn: The noise figures between passive stages can be added together; for example, the loss of the preselector adds directly in decibels to the noise figure of the succeeding amplifier (in decibels). The noise figure of the mixer can never be less than the decibel sum of the conversion loss of the mixer and the noise figure of the first IF amplifier. As a practical matter, the noise figure of the mixer is usually the decibel sum of the IF amplifier noise figure filter loss and mixer conversion loss, since $t = 2$ for most cases. There is also the possibility of oscillator spectrum noise being introduced into the RF port of the mixer, an important consideration in homodyne or DC IF-type receivers that incorporate IF amplifiers that extend to zero frequency.

The noise figure of wideband RF preamplified superheterodyne receivers must be carefully calculated since noise at the image frequency, called the double sideband (DSB), may be present. The single sideband (SSB) noise figure is generally a more desired value. Consider Figure 7-11. In Figure 7-11a, the noise figure will be the double sideband value since in a superheterodyne the mixing action creates two sideband responses from mixing

$$f_{IF} = f_{LO} + f_{RF}$$

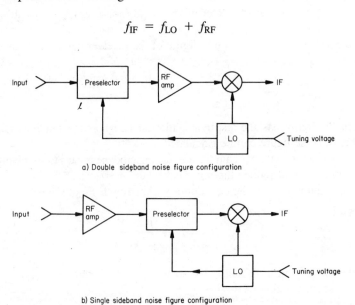

a) Double sideband noise figure configuration

b) Single sideband noise figure configuration

Figure 7-11. (*a*) SSB and (*b*) DSB configurations for an RF amplified preselected superheterodyne. Receiver (*a*) single and (*b*) double.

where

$$f_{IF} = \text{the intermediate output frequency}$$

$$f_{LO} = \text{the local oscillator frequency}$$

$$f_{RF} = \text{the input frequency}$$

or

$$f_{RF1} = f_{LO} + f_{IF}$$

$$f_{RF2} = f_{LO} - f_{IF}$$

Either f_{RF1} or f_{RF2} can be used for the signal (the lowest is most often chosen), but if the noise-limited case (sufficient confining gain) prevails, noise will appear at both frequencies. If the IF and preselector bandwidths are purposefully chosen to reduce image response, the signal will only appear in one bandwidth, whereas the noise will be present in both. This contributes a 3 dB increase in the total noise, suppressing the signal. The preselector loss must also be added (in decibels) to the noise figure of the RF amplifier, degrading the sensitivity even further. The advantage of this arrangement, however, is that only signals present in the narrow preselector bandwidth are fed to the RF amplifier, thus reducing overload or jamming of the entire band by one or more signals outside the narrow RF frequency band of interest. Figure 7-11b, which places the preselector after the RF amplifier, gives a 3 dB improvement due to achieving the single sideband noise figure by the elimination of the image sideband noise and the preselector loss; however, the entire B_r of the RF amplifier is exposed to the RF environment.

The ideal solution is the use of a preselector *before* and *after* the RF amplifier requiring that they be tuned or tracked synchronously (tuned to the same frequency). Yittrium iron garret (YIG) voltage tuned filters are commonly used for this purpose and for generation of the variable local oscillator. Tracking can be readily accomplished with this arrangement since the YIGs are tuned by a magnetic field generated by a tuning current. This current is quite similar for each YIG assembly, facilitating tracking. One filter, the preselector, preceeds the RF amplifier, the other, the post-selector, follows the RF amplifier accomplishing the objectives stated above. (The YIG preselector may also be used for the RF amplified crystal video receiver.)

The mixing process of a superheterodyne gives rise to images and a wide variety of spurious signals. This is due to multiple harmonics of the signal creating mixing products that appear at the IF frequency. Care must be taken to assure spurious free performance in superheterodyne DF channels since both amplitude and phase of matched channels can be affected.

The bandwidth of a superheterodyne receiver is usually dictated by the video bandwidth required for maximum fidelity of the received pulse. There are many conflicting requirements between using the narrowest video bandwidth to optimize detectability at the expense of pulse fidelity and using the widest bandwidth for maximum amplitude fidelity. Passive direction finding is generally driven to the latter choice, since monopulse comparisons require sufficient amplitude to be present long enough to take the ratio. B_v for maximum fidelity is usually established by the rise time t_r of the pulse as follows:

$$B_v = \frac{.35}{t_r} \qquad (7\text{-}22)$$

This is not the value for optimum signal-to-noise ratios, which is

$$B_v = \frac{1}{pw} \qquad (7\text{-}23)$$

The pulse in Equation (7-23) would be detectable as a triangular shaped signal maintaining a peak value equal to its amplitude and an arrival time giving accurate PRI. Equation (7-22) would yield a replica of the input pulse. In passive DF applications, most receivers are attempting to receive signals of unknown pulse characteristics, and use of a wider bandwidth sometimes enhances the probability of intercept at some small expense in signal-to-noise ratio. It is also useful to maintain fidelity for pulse width measurements, which is not feasible using Equation (7-22).

The RF bandwidth is governed, in a preselected superheterodyne, by the choice of the IF and the need to reduce spurious responses. Since the image frequency is always $2f_{\text{IF}}$ away from the real signal frequency, the maximum preselector bandwidth is at least less than $2f_{\text{IF}}$ and, more generally, is

$$B_{\text{RF}} \cong 2B_{\text{IF}}$$

Figure 7-12 shows these general relationships. The information bandwidth necessary to recover the pulse is B_v. The IF bandwidth B_{IF} is larger than B_v, although it is usually chosen to be twice as large; B_r is at least equal to B_{IF} and is often many times its value.

The sensitivity of the superheterodyne receiver may be determined in the same manner as the noise-limited (amplified) crystal video receiver; the gain in this case is provided either as RF gain, IF gain, or a combination of both. The mixing processes in a superheterodyne are linear despite the use of diode devices, making frequency conversion of the RF signal to the IF frequency a matter of simple translation. In early designs when RF gain at microwave frequencies was not attainable, the conversion loss of a good mixer (~ 9 dB) could be utilized by adding IF gain to define the mixer as the predetection noise source compared to the gain-limited postdetection crystal detector. The gain obtained was more readily developed at the IF, since it was a lower frequency. Frequencies of 30, 60, and 100 MHz were popular, the usual compromise being the requirement for wide B_v for pulse fidelity in contrast to low-noise figure of the first IF amplifier stage. (B_v is more

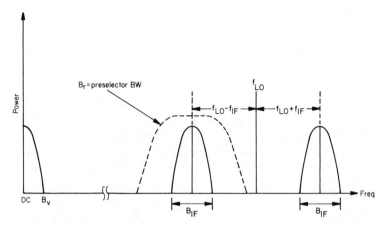

Figure 7-12. Bandwidth relationships in the superheterodyne receiver.

easily obtained at relatively higher IF frequencies since for a fixed bandwidth it is a smaller percentage of the operating frequency.) The noise figure compromise directly reduced the conversion loss of the mixer since the IF noise figure adds directly to the mixer. Receivers of this type are called down-conversion superheterodyne, implying an IF frequency below the operating frequency.

The sensitivity of this type of receiver must consider three bandwidths: the RF bandwidth, B_r, determined by the RF frequency bandpass device and where it is placed (DSB versus SSB); the IF bandwidth, B_I, which is generally made equal to B_v; and the video processing bandwidth.

It is usual practice to assume that the superheterodyne receiver is operating linearly, permitting the expression of sensitivity to relate to the equation of overall noise figure F_t.

The sensitivity of a linear superheterodyne for the case of $B_{IF} = 2B_v$ may be determined reasonably accurately for output signal-to-noise ratios $\gg 0$ dB by using the equivalent B_e Equation (7-8) in conjunction with the noise figure Equation (7-20) as follows:

$$B_e = \sqrt{2B_r B_v - B_v^2}$$

where

$$B_r = \text{the smaller of the IF or RF bandwidths}$$

$$B_v = \text{the video bandwidth}$$

Computing the overall noise figure F_t in decibels yields

$$P_r = -114 + F_t + 10 \log B_e + (S/N)_o$$

where

$$(S/N)_o \gg 0 \text{ dB}$$

P_r = the minimum signal that will provide the desired $(S/N)_o$

This equation is reasonably accurate for the conditions stated above. There are, however, certain cases that limit the above approach: In receivers that have high processing gain it may be possible to operate with output signal-to-noise ratios that are less than 0 dB. There are many applications where the video bandwidth is considerably narrower than the RF bandwidth, leading to the equivalent of processing gain (narrow bandwidth CW receivers, for example).

Under these conditions, it is necessary to go back to the methods of determining the input $(S/N)_I$ ratio for a desired video output $(S/N)_O$ ratio for the various ratios of B_v/B_r as before. To accomplish this, Figure (7-13) is a curve of a linear detector, plotting the bandwidth ratios parametrically relating $(S/N)_I$ to $(S/N)_O$. By examination, it may be seen that when $B_r \cong 2B_v$, $(B_v/B_r = .5)$, the $(S/N)_O \cong (S/N)_I$ for output ratios greater than zero. For output ratios less than zero and for very narrow band video or, conversely, very wide bandwidth IF or RF conditions, there are major differences.

The procedure to follow duplicates that described previously.

1. Determine the required output signal-to-noise ratio.

2. Using the ratio B_v/B_r, read the input signal-to-noise ratio required at the detector.

3. Compute the input signal power P_s at the receiver by using the equation

$$P_s = -114 + F_t + 10 \log B_r + (S/N)_I$$

This is the signal that will give the required output signal-to-noise ratio for the bandwidths involved.

7.2.2 Operational Considerations

Superheterodyne dynamic ranges are consistent with the crystal video detector logarithmic video amplifier (DLVA) approach since a new log–log video amplifier is now used. This will be discussed in Chapter 10; however, it is important to note that modern technology permits the design of a high-quality logarithmic amplifier with wide bandwidths (greater than 40 MHz typically) and the capabilities of providing both RF and demodulated RF to video output for other superheterodyne or crystal video receivers.

Figure 7-14 shows one channel of a typical superheterodyne receiver used by a scanning superheterodyne amplitude monopulse type DF receiver. Two channels would be required for the monopulse system. The aforementioned dual YIG type of preselection is utilized for optimum noise and signal performance by scanning the YIG preselectors with the local oscillator. The receiver is shown with a logarithmic

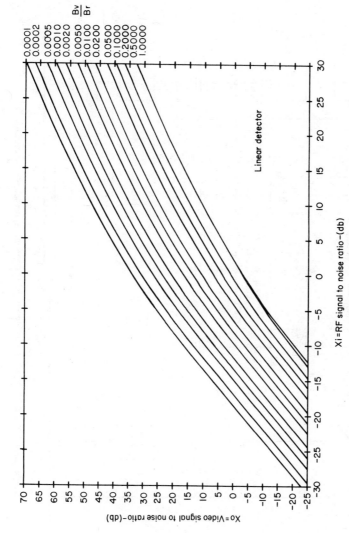

Figure 7-13. RF signal-to-noise ratio as function of video signal-to-noise ratio for various B_v/B_r ratios for a linear detector.

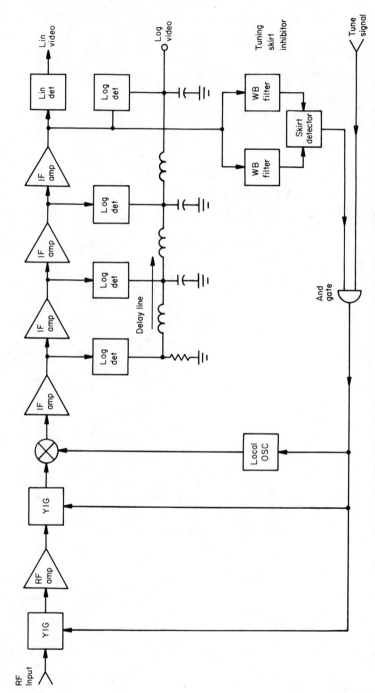

Figure 7-14. Single-channel "skirt inhibited" tuned preselector logarithmic superheterodyne receiver.

amplifier constructed by placing saturating or logarithmic detectors at each stage and adding the resultant video through an artificial delay-line. A weak signal is amplified by all n stages of the IF; a strong signal saturates the latter stages, which act as fixed voltage pedestals upon which the earlier stages add their unsaturated signals; the delay-line compensates for individual stage delays. The output of the delay-line therefore is an arithmetic or voltage sum for an input that is geometric. Doubling the RF input, for example, merely adds a voltage (representing a 3 dB input change) to the output. This may be recognized as the desired logarithmic response necessary to achieve wide dynamic ranges and necessary to form the monopulse ratio by subtraction in a multichannel receiver. It is important to note that measuring the half-voltage video level out of a log system (superheterodyne or DLVA) will *not* give pulse width. The log action requires that a fixed voltage be subtracted, as determined by the decibels-per-volt slope of the logarithmic curve. It must also be noted that the bandwidth of a log IF amplifier will change greatly as a function of signal level since the addition and/or deletion of saturating stages effectively widens the response in much the same fashion as a limiter, which increases the spectrum of the limited signal as its level increases by sharpening the rise time. The number of stages in use in a log IF amplifier is determined by the RF input signal, and since the stages contain reactive elements, the number of bandwidths determining poles will vary accordingly, as each stage saturates.

Most superheterodynes scan across a wide bandwidth (typically an octave) and stop when a signal crosses a threshold, set at a level above the noise to ensure an acceptable false alarm rate, the intent being to place the intercept spectrum within the passband for analysis and to be able to record the local oscillator frequency to use as an identification word for that signal. It is therefore necessary to be assured that the signal is centered within the passband. If the sin x/x pulse spectrum of a very strong pulse crosses the threshold when one of the side lobes is tuned to by the receiver by the skirts of the IF filter ($B_r > B_{IF}$), differentiation of the signal will occur and the leading edge and trailing edges will appear as "rabbit ears" as shown in Figure 7-15b, a condition where "false triggering" has stopped the local oscillator tuning. The receiver is not centered on the signal properly and may be stopped far from the center RF frequency of the pulse. This effect may be overcome by a circuit called a skirt-inhibitor. Returning to Figure 7-14, it may be seen that the (relatively) linear RF output of the amplifier feeds both a wide and narrow bandwidth skirt inhibitor each tuned to the same IF frequency. If the stopping rules of the receiver are properly written, stopping will only occur when both are wide and the narrow band filter thresholds are crossed, as shown in Figure 7-15c. The narrow bandwidth will distort the pulse; however, the receiver will be properly tuned within the criteria. The wide bandwidth in Figure 7-15d can then be used for signal measurements, since the signal is fully centered within it and there is no distortion.

Many different superheterodyne configurations are used. Where YIG or mechanically tuned preselectors are not desired, image-cancellation mixers can be used to prevent stopping at the image frequency or to provide actual cancellation of the image itself. In other cases, the IF frequency can be chosen to range from DC to

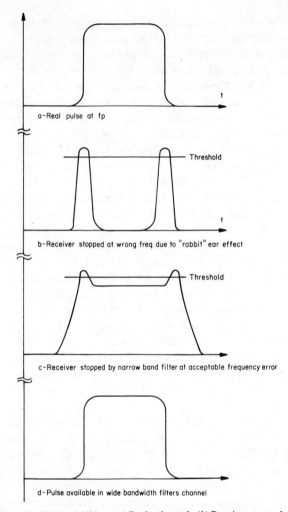

a-Real pulse at fp

b-Receiver stopped at wrong freq due to "rabbit" ear effect

c-Receiver stopped by narrow band filter at acceptable frequency error

d-Pulse available in wide bandwidth filters channel

Figure 7-15. Action of an IF skirt inhibitor. (*a*) Real pulse at f_p. (*b*) Receiver stopped at wrong frequency due to "rabbit ear" effect. (*c*) Receiver stopped by narrow band filter at acceptable level. (*d*) Pulse available in wide bandwidth filters channel.

B_v, allowing reception of either the real or the image frequency to stop the receiver. In this case, the resultant frequency resolution (error = $2B_v$) must not be significant, as is the case for receivers operating at high (>10 GHz) RF frequencies. The advantage here is the improved intercept probability and receiver simplification.

Figure 7-16 shows a homodyne configuration so named for the self-generation of the mixing signal. An oscillator f_o at the IF frequency f_{IF} is mixed in a single sideband modulator (mixer) with one-half of the input signal f_s, generating $f_{LO} = f_s + f_O$. This output, in turn, is mixed with the other half of f_s to generate f_{IF} since $f_{IF} = f_O$. The advantage of the configuration is the potential for wideband, high-

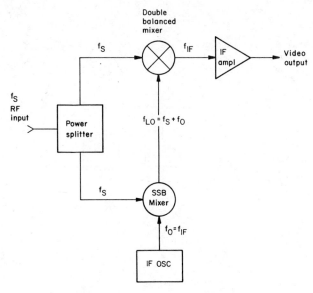

Figure 7-16. Homodyne superheterodyne.

probability detection. The disadvantage is feedthrough of f_{LO} into the IF (since $f_{LO} = f_F$) and the requirement for high spectural purity of the oscillator since its DC to "close in" noise adds directly in the mixer, tending to reduce the overall noise figure.

7.3 DOWNCONVERTER RECEIVER

The advantages of a wide bandwidth crystal video receiver are simplicity, wide bandwidth, high frequency-intercept probability, and low cost. The advantages of the superheterodyne are high sensitivity, high signal strength detection probability, frequency resolution, and a narrower RF bandwidth that may exclude unwanted signals. This combines the advantages of better jamming and strong signal environment immunity with the disadvantage of reduced instantaneous detection probability. When an RF amplifier is added to the crystal video receiver, some of the super-heterodyne features are attained; however, there are added problems of dynamic range and expense.

   In the area of warning receivers, a new configuration called the baseband receiver has appeared. This configuration which is essentially a downconverter technique, attempts to attain the best features of both configurations. Figure 7-17 shows the methodology. The usual 2–18 GHz frequency band is multiplex filtered into four 4 GHz wide RF segments with the purpose of converting each to a 4 GHz baseband in the 2–6 GHz range. The first channel is 2–6 itself and is fed directly to SW3 for transfer to the 2–6 GHz baseband system. The 6–10 GHz and the 14–18 GHz

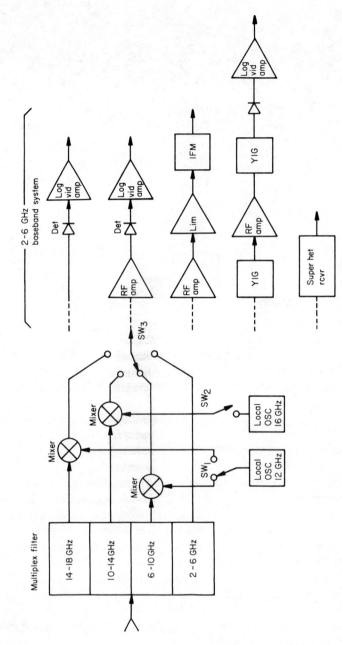

Figure 7-17. Down-converter receiver—single channel shown.

range are mixed with a 12 GHz oscillator to effect 6–2 and 2–6 GHz down conversions, respectively. SW1 permits either channel to be selected, or the switch can be replaced with a power-splitter for simultaneous operation. The 10–14 GHz range is beat with a 16 GHz local oscillator for its 2–6 downconversion.

Once the signals are translated or switched to the 2–6 baseband range, any one or more of the five various types of receivers shown can be used to provide appropriate demodulation. All signals can be superimposed at one time for maximum intercept probability or selectively switched in an a priori sequence dependent upon signal density. Note of the switch positions provides the most significant frequency word with the lesser frequency word developed by one of the techniques.

Since for DF applications, it is generally necessary to have at least two channels operative to form the monopulse ratio, the downconverter system is expensive. A baseband processing system such as an instantaneous frequency receiver, for example, may be assigned after DF is determined by a simple four-channel DF system such as the gain-limited or noise-limited detector. It is also possible to combine RF channels and switch them to resolve ambiguities (17).

The need for higher sensitivities and better capabilities to operate in high-density environments has favored the baseband downconverter utilizing multiple channels as the signal density increases. In almost all cases, these types of systems are DF driven; that is, DF measurement is made on a pulse-by-pulse basis and processed, the frequency capability being assigned to resolve ambiguities and reduce signal densities when necessary. The configuration pictured here is generic and is only shown to describe the concept, since there are serious considerations of spurious products that have not been considered.

The sensitivity of the downconverter can be calculated for the TSS or general output (signal-to-noise) case using a combination of the methods presented above. It is essential to do all system analyses carefully, considering all losses and gains, since any one design may be the combination of two or more techniques for calculation purposes. At this point, the reader should be very familiar with the necessary procedures.

REFERENCES

1. Davenport, W. B., Jr., and W. L. Root, *An Introduction to the Theory of Random Signals and Noise*, New York: McGraw-Hill, 1958, p. 266.
2. Ayer, W. E., "Characteristics of Crystal-Video Receivers Employing R-F Preamplification," Technical Report 150-3, Stanford Electronic Laboratories, Stanford, CA, 1959.
3. "Dynamic Range Extension of Schottky Detectors," Hewlett-Packard Application Note 956-5.
4. "The Hot Carrier Diode Theory, Design and Application," Hewlett-Packard Application Note 907, p. 4.
5. Schottky, B., "Diode Video Detectors," Hewlett-Packard Application Note 923.
6. "Mixer and Detector Diodes," Ref. 80800, Alpha Industries Application Note.
7. Moulo, R. B., and F. M. Schumacher, "Tunnel Diode Detectors," *Microwave Journal*, Jan. 1966, pp. 77 ff.

8. Lucas, W. J., "Tangential Sensitivity of a Detector Video System with RF Preamplification," *Proc. IEE VII3*, No. 8, Aug. 1966, pp. 1321–1330.

9. Uhlir, A., Jr., "Characterization of Crystal Diodes for Low-Lead Microwave Detection," *Microwave Journal*, July 1963, pp. 59 ff.

10. Adam, S., R. Riley, and P. Szente, "Low-Barrier Schottky Diode Detectors," *Microwave Journal*, Feb. 1976, pp. 54 ff.

11. Cowley, A. M., and H. O. Sorensen, "Quantitative Comparison of Solid-State Microwave Detectors," *IEEE Transactions on Microwave Theory and Techniques*, MTT-14, No. 12, Dec. 1966.

12. Rice, S., "Mathematical Analysis of Random Noise," *Bell Systems Technical Journal*, Vols. 23 and 24, 1945.

13. Klipper, H., "Sensitivity of Crystal Video Receivers with RF Preamplification," *Microwave Journal*, Vol. 8, Aug. 1965, pp. 85–92.

14. Tsui, J. B., *Microwave Receivers and Related Components*, NTK Accession No. PB 84-108711, 1983, p. 13, Section 2.3.

15. Lipsky, S. E., "Calculate the Effects of Noise on ECM Receivers," *Microwave*, Oct. 1974, pp. 65 ff.

16. Hughes, R. S., "Determining Maximum Sensitivity and Optimum Maximum Gain for Detector-Video Amplifiers with RF Preamplification," Naval Weapons Center, Technical Memorandum 5357, NWC, China Lake, CA, Mar. 1985, p. 14.

17. Konig, C., "ELINT Design Melds Classic Methods," *Microwaves & RF*, Sept. 1984.

Probability of Detection

The probability of detecting a signal depends upon many factors in the passive DF receiver system. There are the obvious necessities of being tuned to the intercept frequency, to having the intercepting antenna pointing to the intercept, and having sufficient intercept sensitivity to ensure reception of the target signal down into its back lobes to remove the target scan effectively, if possible. If the antennas are rotating and sensitivities are limited, a "beam-on-beam" condition, where both antennas must point in each other's direction, may be the only time intercept is accomplished. If pulses are transmitted, the receiver must be ready when the target is transmitting. Since these conditions prevail for each intercept, it becomes obvious that the control and setting, or adaptability of the passive DF system, are the major factors in successful measurement of threat parameters, especially in dense environments.

To ensure proper operation of complex receiving systems, the concept of computer control immediately suggests itself and indeed is the method most commonly used. This idea has been rapidly advanced by the availability of faster and better digital devices and the obvious acceptance of digital methods in all areas of electronic technology. It has also been advanced for passive direction finding by the use that is made of this information. Since passive DF data derived for threat detection is used for protection, decision making, and control of countermeasure assets, the idea of "power-managed" defensive countermeasures has evolved as the optimum way to direct protective assets to counter priority threats. The electromagnetic environment signal density has increased to the extent where the decision-making processes in power-managed assets must also be made rapidly. This computer-aided operator decision-making or semiautomatic operation is becoming the standard. As the signal density further increases, fully automatic decision making, aided by artificial intelligence concepts, will undoubtedly prevail.

The problem then is one of making the correct choice. Is there an intercept or not? Is the system being jammed to make it overwork to reduce the threat of probability of detection? Is the equipment presenting false alarms, each of which occupies time during which other events can occur? To address these questions, it is necessary to consider the first and perhaps most significant factor in a passive DF system: the false alarm rate (FAR). This is a number representing the number of false alarms occurring in a given period of time and is a function of the setting of the receiver's threshold-of-detection.

8.1   FALSE ALARM PROBABILITY

In passive DF systems, it is becoming more important to utilize the angle-of-arrival as a sorting tool since frequency-driven receivers are problematic in the presence of noise, frequency dispersive signals, dense environments, and other high dutycycle situations. DF sorting is also important since the angle-of-arrival of a signal cannot change as quickly as its frequency, especially for more distant targets that can be detected by more sensitive receivers.

Most receivers, especially those used for warning, integrate over many pulses to improve the sensitivity. This is especially true in radar, where the CRT screen persistence is used to improve bearing accuracy. In automatic systems correlation, integration and averaging accomplish the same purpose. Monopulse radar, in fact, is generally not "mono" at all; its chief advantage being the signal normalization function provided by the method, not the single-pulse detection capability. With the need for DF driven ESM systems, however, this has changed, and detection and determination of bearing on a single pulse has become a necessity since sorting is done on a pulse-by-pulse basis. To do this, a decision-making process must take place. For the DF systems we have studied, this decision is the crossing of a voltage threshold set at a level above the noise floor, representing the number of decibels above random noise that a voltage, signal, or noise must attain to assure a certain false alarm rate. The setting of this threshold is an important factor in establishing the receiver operating sensitivity, since the higher it is set to reduce false alarms, the greater the signal amplitude necessary to be recognized as a signal to be processed. A low threshold will allow processing of weaker signals at the penalty of false alarms, which may or may not be useful depending upon the capability of the system to average over many crossings. What is needed is a high enough threshold optimized to accomplish conflicting objectives. How do we choose it?

To set practical limits, we must tolerate a certain number of false alarms. This is a number that is selected as part of the system design. For example, if we are going to launch an atomic missile we would want no false alarms; our threshold would certainly be high. If we are controlling a less potent system, we would perhaps tolerate one false alarm per hour or even one false alarm-per-second. The tolerance factor or false alarm rate is system dependent and all other things being equal, usually a function of the system's computer capacity since an extensive capacity for storage and processing allows more false alarms to be tolerated than a limited system that has little memory or a long processing time. Excess computer capacity can be used to average signals, if it is available.

8.1.1   Envelope Detector

To consider the probability of a burst of noise occurring above a threshold, instead of a signal, it is necessary to evaluate many conditions relating to the detector. Is it an envelope detector where the video bandwidth B_v is optimized for the best pulse detectability or is it a square-law or linear device? What type of amplifier follows the detector: linear or logarithmic? Is a signal present with the noise or is it only the noise peak above the noise floor that is triggering the system? As can be imagined, the analysis of all of these factors is difficult, and since much of the theory is statistical, most system designers view the results skeptically. The problem relates to the many variations stated above and the fact that formulation for the established radar case does not always apply to the passive DF receiver. To study the effects of noise, it is necessary to analyze the critical case of noise alone present in a thresholded system.

Consider first the classical radar detection case as stated by Skolnik (1). This is the case of a superheterodyne radar receiver with sufficient gain to be noise-limited and with an RF bandwidth determined by the intermediate amplifier bandwidth such that

$$B_r = B_{IF} \cong 2B_v$$

where

$B_r = B_{IF}$ is the RF noise bandwidth as determined by the IF bandpass

$B_v =$ the video bandwidth

This type of detector is referred to as an envelope detector. It extracts the shape (envelope) of the modulating signal (pulse) by passing the bandwidth B_v of the informational signal (radar return) while rejecting the RF or IF carrier completely. Since this analysis has been used most frequently it will be considered here; however, envelope detection is rarely encountered in the passive DF receiver since the width or bandwidth occupancy of the pulse to be detected is not usually known and B_r is made much greater than B_v by design to improve intercept probability. Figure 8-1 is a diagram of the detection model. Noise entering the system is assumed to

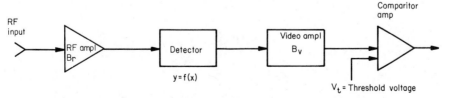

Figure 8-1. Detector model for noise-limited false alarm determination.

be white and of a Gaussian shape, which means that it follows the noise probability density function

$$P(v) = \frac{1}{\sqrt{2\pi}} \exp\left(-\frac{v^2}{2\sigma^2}\right)$$

where

$$\sigma^2 = \text{the noise variance}$$

$$V = \text{the noise voltage}$$

This function is the probability or chance of finding the noise voltage between V and $V + d(v)$. The mean value of v is zero. When this Gaussian noise is passed through a narrow-band filter and is envelope detected, its characteristics are changed from a Gaussian distribution to a Rayleigh distribution of the form

$$P(v) = \frac{V}{\sigma^2} \exp\left(\frac{-V^2}{2\sigma^2}\right) \qquad \text{for } V \geq 0$$

$$P(v) = 0 \qquad \text{for } V \leq 0$$

This equation is of the form of a Rayleigh distribution. It is possible to plot a curve of the average time between false alarms, defined as TFA, as a function of the threshold voltage V_t to RMS noise voltage for various values of B_v, the video bandwidth. Skolnik defines a term $\Psi = \sigma^2$ as the RMS noise power. A similar curve is drawn in Figure 8-2. For use with the values usually encountered in passive DF systems, the curve has been expanded and plotted as the log of seconds versus the threshold-to-noise log voltage to permit greater accuracy. The equation plotted is a plot of

$$\text{TFA} = \frac{1}{B_{RF}} \exp(V_t^2/2\sigma^2) \qquad (8\text{-}1)$$

For example, in a 10 MHz video bandwidth system for TFA $= 100$ seconds (log $100 = 2$), the V_t-to-noise threshold should be set to 13.13 dB. As an approximation PFA, the probability of false alarm for the narrow RF bandwidth case

$$\text{PFA} \cong \frac{1}{2(\text{TFA})(B_v)} \cong \frac{\text{FAR}}{2B_v} \qquad (8\text{-}2)$$

since

$$\text{FAR} = \frac{1}{\text{time between false alarms}}$$

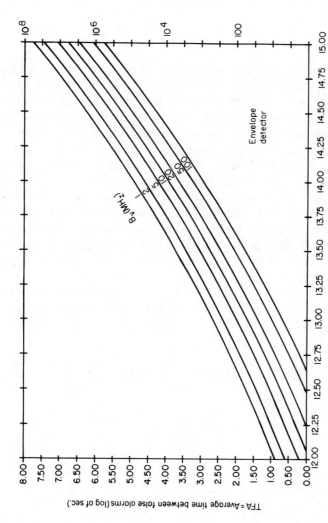

Figure 8-2. Average time between false alarms as a function of video bandwidth for various threshold-to-noise levels for envelope detection.

211

This approximation assumes that the average duration of a noise pulse is equal to one-half the inverse of the video bandwidth.

8.1.2   False Alarm Probability for Wide RF Bandwidths

Equation (8-1) does not consider the case most often encountered in most warning receivers; that is, that of a wideband RF, $B_r >> B_v$. It also does not address the wideband square-law detector or the use of logarithmic amplifiers ahead of the threshold process. Several attempts have been made to account for these differences. When a wide RF bandwidth of noise is essentially folded into a narrower video bandwidth, a smoothing filter effect takes place. For this reason, it is not easy to predict the false alarm rate reliably. The matter has been considered by Skolnik (1), Harp (2), and Tsui (3). The probability density function of quadratic-detected wideband RF noise, filtered by a narrow-band video filter, is of the form of a chi-square function. For a linear detector, with $B_r >> B_V$, the probability density function is assumed to be half-Gaussian. These functions are shown in Figure 8-3 for each of the three cases. It should be noted that in each case the total area under the curve is unity ($P = 1$), the significance of which is that at some level threshold crossing will always take place.

To determine a method of analysis for the wide bandwidth, quadratic case, the probability of false alarm (PFA) for the square-law detector must be calculated from the error function (erf) as follows:

$$\text{PFA} = \int_{V_t}^{\infty} P(v)d_v = 1 - 2 \text{ erf} \left(\sqrt{V_t/\sigma^2} \right) \tag{8-3}$$

where

$$\text{erf} = 1/\sqrt{2\pi} \int_0^x e^{-y^2/2} \, dy \tag{8-4}$$

substituting (8-4) in (8-3) we get

$$P(V) = \frac{1}{\sigma\sqrt{2\pi}} \exp(-V/2\sigma^2) \tag{8-5}$$

for $V > 0$ for the chi-square case.

The FAR is related to the PFA as in Equation (8-2). The exact values can be calculated from error function tables (see Ref. 4). An easier method is to use Figure 8-4, which is a plot of the time between false alarms as a function of B_v, assuming B_r is $> 20 B_v$. This plot permits determination of the false alarm rate.

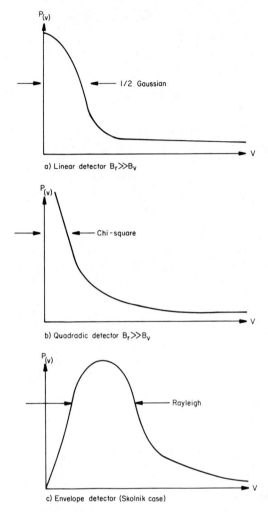

a) Linear detector $B_r \gg B_v$

b) Quadradic detector $B_r \gg B_v$

c) Envelope detector (Skolnik case)

Figure 8-3. Probability density function. (*a*) Linear detector $B_r >> B_v$. (*b*) Quadratic detector $B_r >> B_v$. (*c*) Envelope detector (Skolnik case).

8.2 PROBABILITY OF SIGNAL DETECTION IN NOISE

Up to this point we have determined the false alarm rate of a system by considering the possibility of a noise-crossing of a threshold set a certain number of decibels above RMS noise. We have considered the postdetection case and have related the problem to the three cases commonly encountered, namely, the envelope-detector or radar case, the wide RF bandwidth wide-open receiver with quadratic-detection, and the intermediate case of a scanning receiver with wide RF bandwidths and

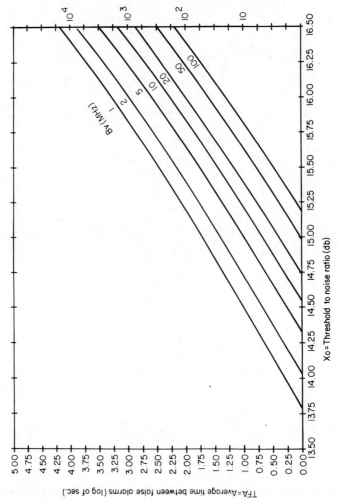

Figure 8-4. Average time between false alarms for $B_R >> B_V$ for a quadratic detector.

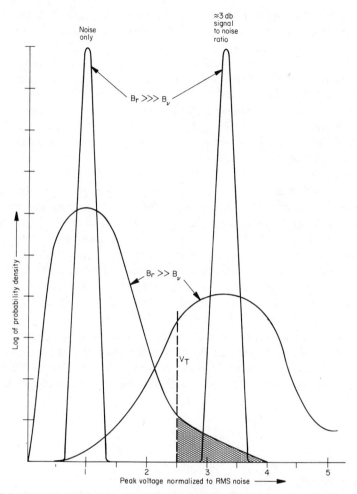

Figure 8-5. Output probability signal densities for Gaussian noise at the output of a linear detector.

video bandwidths not necessarily optimized for the input signal. We must now determine the probability of detecting (P_D) a *signal* in the presence of noise.

When a signal is added to random noise, the presence of both effectively causes a shift in the probability density function as compared to noise alone. This is shown in Figure 8-5, which is a plot of the log probability density as a function of normalized threshold peak-to-RMS noise for the Gaussian input case for a linear detector. The predetection curve for noise only, which has a zero mean value (centered about the probability density axis), is translated into a new probability density function shifted to the right by the addition of the signal. The area of the curve is a constant, but the height varies as the RF bandwidth increases with respect to the video bandwidth. When a signal is added, the probability density function is shifted due to the addition of the noise on top of the signal. The threshold V_t shown in the figure will be crossed by noise due to the probability density function with no signal present. When a signal is added and the probability density function is shifted to the right, the FAR threshold crossings due to noise are represented by the crosshatched area. Crossings, due to the signal in the presence of noise, in the rest of the area to the right of the crosshatched threshold area represent the probability of detecting the signal. In passive direction-finding for pulse signals, if a video bandwidth is chosen to be wide for minimum width pulses, wide pulses will further change the P_D due to a greater look window resulting from the wider pulse and the fact that the pulse will reach its peak value.

The probability of detecting a signal may be the driving factor and will impose a higher threshold requiring a signal level many decibels above the threshold set for the FAR. To illustrate this, Figures 8-6 to 8-9 have been plotted to provide the probability of detecting a signal as a function of the signal-to-noise ratio selected (for whatever the criteria). Although these curves are for a linear detector, they may be used with reasonable accuracy for a square-law detector as well, since the signal level is high enough to predominate. The threshold-to-noise ratio is the setting that may have been established by the FAR or by limitations in the receiver. The curves show the actual signal-to-noise ratios that will be necessary for a linear detector to achieve the probability of detection values (P_D) listed.

The first curve, Figure 8-6, assumes that B_r is chosen to be wide enough for the *minimum* expected pulsewidth and that this is the pulse that is received. If, for example, we had set the threshold above noise by 13.7 dB, we would need an actual 19.75 dB signal-to-noise ratio to achieve 0.90 probability of detection. We assume that

$$\text{MIN PW} = \frac{1}{B_v}$$

where B_v is measured at the 3 dB point.

It is obvious that if 13.7 dB were chosen for an adequate FAR, the receiver would have to obtain the higher 19.75 dB value if 0.90 probability of detection were to be obtained. It would also be necessary to check to see if 19.75 dB was high enough to obtain the required accuracy.

Figures 8-7, 8-8, and 8-9 show that given the choice of B_v, if a pulse two, four, or eight times as wide as the minimum were received, the signal-to-noise requirements would drop significantly and could approach the signal-to-noise ratio chosen for the FAR alone. It is not to be concluded from these curves that the wider B_v is made, the better, since widening B_v will increase the noise FAR, necessitating new thresholding. Computer programmed variable thresholds are often used to achieve constant FAR systems, with programming made variable and appropriate to specific threat types. Since most systems integrate signals, the high signal-to-noise ratios stated here are markedly reduced in practice by integration or the addition of processing gain. Reference 5 gives the effect of pulse integration for the radar case. Interestingly, there is an approximately 5 dB improvement in sensitivity for incoherent

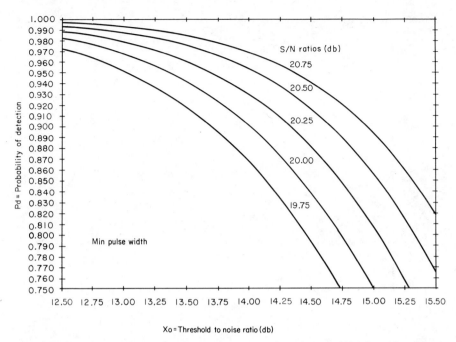

Figure 8-6. Probability of detection versus signal-to-noise ratio for video bandwidth selected for minimum pulse width.

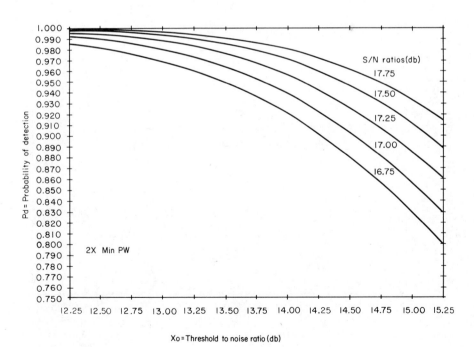

Figure 8-7. Probability of detection versus signal-to-noise ratio of a video amplifier that is two times wider than required for minimum pulse width.

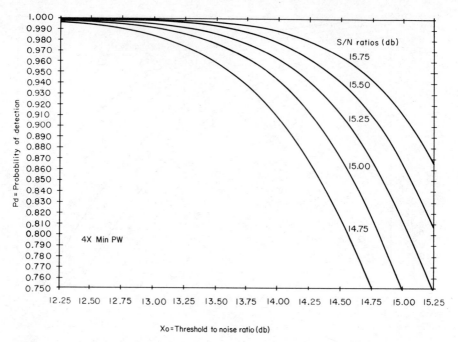

Figure 8-8. Probability of detection versus signal-to-noise ratio of a video amplifier that is four times wider than required for minimum pulse width.

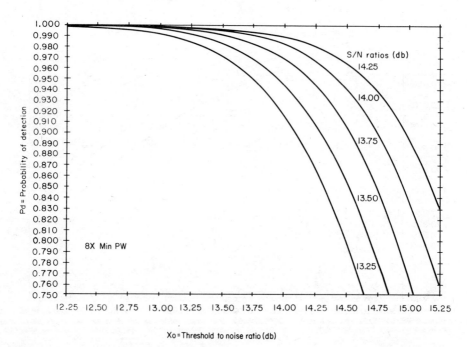

Figure 8-9. Probability of detection versus signal-to-noise ratio of a video amplifier that is eight times wider than required for minimum pulse width.

218

(power added) and 10 dB improvement for coherent (voltage added) pulses for every decade quantity more pulses used. Curves for the quadratic detector are almost identical to the linear-detector shown, allowing both cases to use the same values from the curves.

The important point is that a receiver dedicated to receiving a wide range of pulsewidths necessitating a large B_v will exhibit a greater detection probability for pulses wider than the minimum width the system was designed to receive. The wider B_v will increase the FAR, however, placing a practical bound upon the system. Integration is used, in most receivers, typically to reduce the signal-to-noise ratios to familiar values of 8–15 dB for 2–5 degrees accurate DF systems using wide beamwidth (70–90 degree) antennas.

The method of determining the signal-to-noise ratios and the sensitivity of a noise-limited square-law detector for a given signal-to-noise ratio is our ultimate purpose. We have seen that this is easily done for the envelope-detector case of a narrow RF bandwidth relative to the video bandwidth. It is not as easily done for the wide RF bandwidth case, however; at this point all the methods necessary have been described. To illustrate the technique a step-by-step example will be given.

The receiver to be designed as shown in Figure 8-10 has the following characteristics:

B_v = 1 MHz

B_r = 10,000 MHz
T_{min} = 1 microsecond pulsewidth

TFA = 50 seconds (the time between false alarms)

N = 7 dB (the noise figure of the RF amplifier, which is assumed to have just enough gain to present the noise limited case and a bandwidth B_r)

L = 3 dB (This is the loss of a band-definition preselector of B_r = 10 GHz. It is assumed that the B_r of the amplifier is equal to the filter.)

Steps 1–5 determine the sensitivity for a 0.9 probability of detection:

Step 1. Find X_o from TFA from the curve shown in Figure 8-4 for the quadratic detector.

Step 2. Log TFA = log 50 = 1.7, read X_o for B_v = 1 from curve X_o = 15 dB.

Step 3. Determine the ratio B_v/B_R:

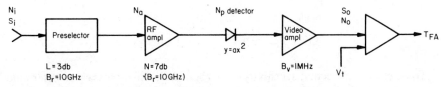

Figure 8-10. Model for a wide RF bandwidth (B_r) narrow video bandwidth (B_v) quadratic detector receiver.

$$\frac{B_v}{B_R} = \frac{1 \times 10^6}{10 \times 10^9} = 10^{-4}$$

Step 4. Use Figure 7-9 to obtain the input signal-to-noise ratio, X_i for $X_o = 15$ dB, read $X_i = -10$ dB for the 0.0001 curve.

Step 5. Compute the noise power (N_p) into the detector from the amplifier (assuming noise limiting $G_A > T_{SS}$) in decibels and taking the noise contribution of B_R/B_v:

$$N_p = -114 + 10 \log(10,000) + N$$

$$= -114 + 40 + 7$$

$$N_p = -67 \text{ dB}$$

Step 6. Determine the input signal S_i from N_i by use of the value X_i from step 4:

$$S_i = N_i + X_i$$

$$= -67 + (-10)$$

$$S_i = -77 \text{ dBm}$$

Step 7. Add 3 dB for the preselector loss

$$S_i = -74 \text{ dBm}$$

This sensitivity assumes an RC roll-off video amplifier that gives an output video waveform

$$V = V_{pk}[1 - \exp -(t/T_{min})]$$

where

t = the pulsewidth (pw)

t_{min} = $1/B_v$ where B_v is the 3 dB video bandwidth

V = video filter output voltage

V_{pk} = peak voltage of the detected pulse for $t = \infty$

V = $0.632 V_{pk}$ for $t = T_{min} = 1$ microsecond

To determine the actual sensitivity for a P_D of 0.9, it is necessary to determine the sensitivities for the various video bandwidths that may be selected for the receiver as part of the design. The -74 dBm sensitivity was determined for the FAR.

Step 8. Use Figure 8-6; going back to step 2, the output signal-to-noise ratio for the FAR is 15 dB, which intersects a horizontal line drawn from the desired detection probability (P_D) of 0.9 at the 20.75 dB signal-to-noise ratio curve. This means that the actual signal input power S_{ip} for P_D of 0.9 is

$$S_{ip} = -74 + (20.75 - 15.0)$$

$$= -68.25 \text{ dBm}$$

for a B_v of

$$B_v = \frac{1}{pw} = \frac{1}{1 \times 10^{-6}} = 1 \text{ MHz}$$

for this case.

If it were necessary to use a video bandwidth twice B_v (2 MHz) Figure 8-7 would give a correction factor of

$$17.35 - 15 = 2.35 \text{ dB}$$

to the -74 dBm sensitivity for a S_{ip} of -71.65. (In this case, interpolation between the curves is required.) The curves are presented for up to eight times B_v and for the values of signal-to-noise ratios usually encountered. It is important to note that the sensitivity is improving for the same intercept as B_v is made wider than is needed for a 1 microsecond pulse. The FAR must be rechecked, however, for large differences. The procedure is general, however, and can be used for other conditions.

Although the above method seems laborious, it relates all of the factors of bandwidths, detector law, and desired false-alarm rate. It is important to realize that wide RF bandwidth systems do not necessarily degrade badly in sensitivity for a given P_D. Comparing the use of this method in the previous example shown in Section 8.1.1, in a 10 MHz B_v for TFA = 100 seconds the signal-to-noise ratio is about 15.85 dB compared to 13.13 calculated previously. Harp (2) in another example compares a narrow bandwidth superheterodyne (B_r = 20 MHz) to a wide bandwidth IFM receiver (B_r = 2000 MHz) for the same conditions of probability of detection, FAR, and receiver noise figure characteristics, finding only a 5 dB improvement in sensitivity for the narrow bandwidth case despite a 100 times increase in the wideband receiver.

The example shown here is for a linear video amplifier and a quadratic-detector assuming sufficient gain for the noise-limited case. Any extra gain will reduce the dynamic range of the detector by moving it into its linear range. At low signal levels, however, superheterodyne detectors often operate in the square-law region with poor signal-plus-noise-to-noise ratios. This is important in analog systems using displays since it is possible to use amplitude versus frequency raster scan on an oscilloscope and pulse integration to detect weak signals in the presence of noise for this method of detection. For digital DF purposes, however, a thresholded signal is usually required, making single pulse crossings important. The equivalent analysis of a linear system can be undertaken by using Figure 8-4, which is the noise-only plot of the average time between false alarms for a linear and approximately quadratic detector in conjunction with Figure 7-13 for the linear-detector. The signal-to-noise input/output ratio curves now must be chosen for either linear- or quadratic-detection as appropriate.

8.2.1 Wide RF Bandwidth FARs Using Log Amplifiers

To account for the use of a logarithmic amplifier following the detection process, it is necessary to consider the compression effects. Ref. (6) discusses the case of a Gaussian noise input with square-law detection followed by a logarithmic amplifier. The logarithmic threshold in millivolts is within the log curve and is stated here as follows:

$$V_T = V_{O\min} + K'[3.622 + P_{in} - P_{in\min} + 10 \log(-\log(PFA)]$$

$$(8\text{-}6)$$

where

$$V_T = \text{log amplifier output threshold in millivolts}$$

$$K' = (V_{O_{\max}} - V_{O\min})/(P_{in_{\max}} - P_{in_{\min}}) \, , \quad (8\text{-}7)$$

which is the voltage output variation of the logarithmic amplifier over the defined power dynamic range of operation (usually given as the slope of the log curve)

$$\text{PFA} = \text{probability of false alarm} = 10^{-6}/\overline{B}(\text{TFA})$$

$$\overline{B} = \text{an equivalent effective bandwidth } B_R \sqrt{2r - r^2}, \text{ where } r = B_v/B_r$$

For a given signal-to-noise threshold not in the log portion, the compression factor K' above can be used to modify the $(S/N)_o$ ratio of strong signals in the log amplifier as long as the noise threshold is in the linear portion of the logarithmic curve. The values of $P_{in_{\min}}$ and the $P_{in_{\max}}$ will correspond to N_o and S_o in (8-7) from which V_T may be determined assuming K' is given for the amplifier. This technique permits determination of the log threshold setting for the $B_R \gg B_v$ case, while Equation (8-6) is the threshold setting for the envelope-detector case ($B_R = 2B_v$).

In the case of logarithmic amplifiers, it must be recognized that in the DLVA the detector is assumed to be in the square-law region and the dynamic range is often obtained by a piecewise linear approximation of square-law operation at high signal levels. In the superheterodyne case, a log intermediate frequency amplifier is used with detectors that usually operate in the square-law region for small signals and in the linear region for large signals.

8.3 INTERCEPT PROBABILITY

Intercept probability is a time coincidence phenomenon as compared to detection or false alarm probability, which depends upon receiver sensitivity, bandwidth, gain, and noise characteristics. Consider a rotating antenna, scanning bandwidth superheterodyne receiver: The sensitivity of such a combination may be high, permitting reception of more than just the mainlobe of the radar transmitter. In the ideal case, the first sidelobes and the backlobe may be received, but the superheterodyne, being a narrowband device, may not have scanned to the frequency of the radar, hence no reception will take place until it does. If the frequency scanning is increased to a high speed such that it is always assured of receiving the intercept frequency wherever the DF receive antenna is pointed to the radar intercept, then the receiver dwell period may be insufficient to receive enough pulses to do integration, and high single-pulse sensitivities will prevail, making only mainbeam detection possible. If both the target and receiver antennas are scanning, it may be possible to attain a beam-on-beam condition where each antenna points to each other. This coincidence will improve detection and, in fact, is almost always required for rotary systems. Maximum detection generally prompts high-speed rotation of the intercept antenna. This will result in detection eventually; however, if the rotations are in the same direction and if the rotation rates are equal, no intercept will occur, this latter condition being generally a trivial case since it is unlikely that both the radar transmitter and radar receiver antennas will be synchronized. Problems can also exist at harmonic multiples of the rotational rates.

The case for rotating antenna systems usually depends upon having time to perform integration as a natural trade-off for system cost and complexity. Although it could be argued that nonrotating systems can essentially remove DF antenna intercept probability problems, the argument becomes less convincing as millimeter wave radars debut. High atmospheric absorption at certain millimeter frequency bands and difficulties in obtaining power and receiver sensitivities at these frequencies make the use of mechanically positioned systems desirable since high antenna gain is more easily obtained with narrow beamwidths. Azimuthal rotation of 360 degrees is not always required, since millimeter radars are usually sector-scanned. DF detecting systems can use the same sector or raster scan techniques using the concept of beam-steering. It may be concluded from this that while phased array and omniazimuthal techniques suggest themselves, mechanical scanning is still a viable proposition.

Frequency scanning also remains an important technique in interception of targets. This is due to high sensitivities and the useful properties of step-scan and look-for prioritization of known frequency ranges. The probability of intercept associated with this process is similar to that of the DF scan. To this must be added the intercept radar pulse time characteristics, since this is another time probability factor.

The current method of expressing intercept probability is to follow the "window concept." This theory, presented by Ref. 7, corrected by Ref. 8, and re-presented by Self (9), represents a reasonably satisfying mathematical mode; however, all characteristics of the system must be idealized, that is, the antennas must have square beamwidths, all receiver bandpasses are ideal, all rates are constant, and so on.

This book will present both methods for evaluation by the reader. In the interests of understandability, the classical methods of determining intercept probability for the antenna beam-on-beam technique and the receive scan intercept probabilty will be considered, first, followed by a discussion of the window concept for generality.

8.3.1   Rotating DF Intercept Probability

Up to this point, for passive direction finding we have assumed that the radar intercept to be received is either constantly illuminating the receiver or that the receiver is a high-sensitivity monopulse design capable of making a DF measurement on each received pulse throughout the radar's scan as the result of normalization or removal of the intercept antenna rotation-rate cancellation. This may be accomplished in a dual-channel ratio circuit, the dual-channel being provided by either another co-boresighted antenna or by an omniazimuthal channel conveniently used for side- and backlobe inhibition of the receiving antenna. The more general case to be considered here will be that of the problem of detecting a *scanning target radar* by a single-channel *scanning DF antenna system*.

Assume the intercept radar antenna has an antenna beamwidth Ψ_r measured 10 dB down from the beam peak and rotates with an antenna scan period of t_r. Assume further that our passive DF system has an antenna beamwidth Ψ_d and a rotational period of t_d, which is less than t_r. It is also implicity assumed that $t_r \neq t_d$ since if they were equal and both antennas rotated in the same direction, it could be argued that no intercept would ever occur (Ref. 8). The probability of detecting a signal in *one scan* of the radar antenna, assuming we are at the proper frequency, is

$$P_D = \frac{\Psi_r}{360} \frac{t_r}{t_d}$$

$$t_d < t_r$$

$$t_d > \frac{\Psi_r t_r}{360} \tag{8-9}$$

where

$$\Psi_r = \text{the 10 dB radar main-lobe beamwidth}$$

$$t_r = \text{the radar intercept's scan period}$$

$$t_d = \text{the DF antenna scan period}$$

The scan periods can be expressed as a function of rotation rate in seconds-per-degree:

$$t_r = \frac{60}{N_r}$$

$$N_r = \text{revolutions per minute (rpm)}$$

The probability of detection for n scans of the antenna is

$$P_T = \sum_1^n [(P + (1 - P)P + (1 - P)^2P + \cdots + (1 - P)^nP]$$

which is a converging geometric series

$$P_D = 1 - (1 - P)^n \tag{8-10}$$

The probability of intercepting the radar signal within a given look time T (noting that n is the total look time T divided by the radar scan time, or $n = T/t_r$) is obtained by substituting Equation (8-9) into Equation (8-10). The inverse of n (which is t_r/T) is called a window. Performing the substitution gives

$$P_D = 1 - \left(1 - \frac{\Psi_r t_r}{360 t_d}\right)^{T/t_r} \tag{8-11}$$

Equation (8-10) can be rewritten in the form

$$P_D = 1 - (1 - P^{-1/P})^{-nP}$$

Since P will be small in one scan,

$$P_D = 1 - e^{-nP}$$

since

$$(1 + P)^{1/P} = e \quad \text{(approximately)}$$

$$P_D = 1 - \exp -(\Psi_r T/360 t_d) \tag{8-12}$$

From Equation (8-12) it may be seen that the probability of detection in a given look time T approaches one as t_d, the period of the DF receiver antenna, approaches zero, which says that the faster the search antenna rotates the better. This rate is limited practically by mechanical problems (speeds up to 1200 rpm have been used) and by the number of pulses it is required to receive from the radar since it is conceivable that the rotation could be faster than a pulse interval. This requires that

$$\frac{t_d \Psi_d}{360} \geq (\text{PRI})_r N \tag{8-13}$$

where

$$\text{PRI} \;=\; 1/\text{PRF of the radar}$$

$$N \;\;\;=\; \text{number of pulses required}$$

In general, rotating antennas with 4 degree 3 dB beamwidth are weighted by a factor of 1.8 to get an equivalent of a 10 dB beamwidth due to the conclusions of Equation (8-13). The usual technique is to consider the 50% probability case and compute t_d based upon the number of pulses desired in Equation (8-13). It will be found that rpms of 1000 are required for single-pulse detection capability making rotating antennas useful for integrating five or more pulses. For this reason rotary systems are usually used with CRT or other integrating systems. Ref. 10 details the use of rotary DF receivers showing that these systems can be made highly competitive where simplicity and cost reduction are essential.

8.3.2 Frequency Scan Probability

The previous section defined the probability-of-intercept of receiving a signal with both the receiver and radar antennas rotating and the receiver tuned to the radar's frequency. This probability was defined as P_D. In this section, we shall assume a beam-on-beam case where both antennas are pointing toward each other, the radar is operating at a constant frequency, and the DF receiver is scanning. In a practical case, this situation is represented by a fixed-antenna system coupled to a high-sensitivity scanning superheterodyne receiver that has sufficient capability to receive the effects of the radar's antenna scan. It is also assumed that the radar is constantly illuminating the area by operating in the monopulse or CW tracking mode.

Figure 8-11 is a diagram of the situation. A radar with a pulse repetition frequency of f_r emits a constant amplitude stream of pulses of width t_w as shown at the top of the figure. The pulses occupy an RF bandwidth (frequency occupancy) of B_R representing the $\sin x/x$ spectrum of the radar pulses.

The passive detecting receiver with an effective bandwidth B_e scans a sector of frequency B that lies between a lower frequency limit f_L and a higher frequency limit f_H. This is shown in the lower frequency versus time plot for the receiver. The receiver is scanning periodically from f_L to f_H at a scan period of t_s.

The probability of detecting a radar pulse in one frequency scan P_{fs} is

$$P_{fs} \;=\; (t_i \;+\; t_w)f_s \tag{8-14}$$

where

t_i = the interval in which the receiver could pick up a radar pulse of width t_w

f_s = the pulse repetition frequency of the radar

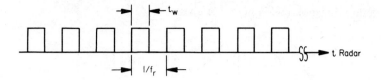

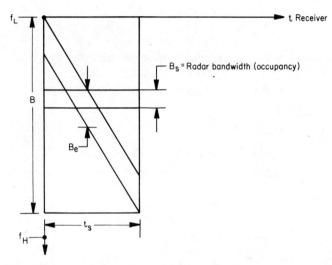

Figure 8-11. Relationships for a nonscanning antenna (constant illuminating) radar as received by a scanning frequency receiver.

The interval t_i is

$$t_i = \left(\frac{B_e + B_r}{B} \right) t_s \qquad (8\text{-}15)$$

where

t_s = the receiver scan rate

B_e = the receiver effective bandwidth

B = the sector of scan between the upper frequency limit f_H and the lower limit f_L

B_r = radar occupany bandwidth

Substituting Equation (8-15) in (8-14) gives

$$P_{fs} = \left[\left(\frac{B_e + B_r}{B} \right) t_s + t_w \right] f_s$$

But since $[(B_e + B_r)/B)]\, t_s$ is greater than t_w for narrow pulse widths,

$$P_{fs} = t_s f_s \left(\frac{B_e + B_r}{B} \right) \qquad (8\text{-}16)$$

The probability of intercepting at least one pulse out of n is

$$P_D = 1 - \text{(probability of missing all pulses)}$$

or

$$P_D = [1 - (1 - P_{fs})^n]$$

if

$$P_{fs} \ll 1$$

then

$$P_D = nP_s$$

since $(1 - e)^n \approx 1 - n_e$ for small values of e

$$P_{fs} \cong nf_s t_s \left(\frac{B_e + B_r}{B} \right) \qquad (8\text{-}17)$$

which is the desired expression for receiving one pulse in n frequency scans for the beam-on-beam condition of a superheterodyne scanning receiver intercepting a pulse signal.

For a CW nonscanning signal, the pulse probabilities (or duty cycle) is 100%, and the receiver scan speed $(1/t_s)$ can be increased; but when the scan is increased beyond a certain speed for a given bandwidth, B_e, there is a loss of sensitivity in the receiver as described in Ref. 11:

$$a_s = \left[1 + 0.195 \left(\frac{B}{T_i B_e^2} \right)^2 \right]^{-1/4} \qquad (8\text{-}18)$$

where

a_s = the response relative to that obtained with a zero sweep rate

B_e = 3 dB receiver bandwidth

B = the sweep width or dispersion (in cycles per second)

T_i = the sweep interval in seconds

If R is the resolving power (equals B_e for $a_s = 1$), then the loss in resolving power is R/B_e:

$$\frac{R}{B_e} = \frac{1}{a_s^2} \tag{8-19}$$

where R/B_e is the apparent bandwidth while sweeping relative to the steady-state bandwidth.

In high-scan speed applications, a scanning receiver that can cover a frequency range during the period of time equal to the pulse width can attain 100% probability of scan detection. Receivers of this type are called compressive or microscan receivers. Originally introduced by White and Saffitz (12) and Kinchloe (13), the compressive receiver holds promise for detection of frequency dispersive signals due to its high P_S. Reference (14) is a recent description of this technology.

8.4   WINDOW FUNCTION PROBABILITY CONCEPT

The probability functions we have been considering are actually all idealized; that is, we are assuming a square antenna beamwidth rotating over 360 degrees in a linear periodic manner, a square receiver bandwidth scanning linearly across a frequency range, and so on. These probability-of-detection functions can be analyzed in the same manner (compare the analysis of the DF scan probability to the frequency scan probability above). The method of doing this was discussed by Hatcher in Ref. 8. Consider a scanning function to be a time slot periodically appearing at a repetitive interval, much in the manner of a pulse train. This is shown in Figure 8-12. Two independent time window functions, $f(A)$ and $f(B)$, represented by pulse trains of widths t_1 and t_2 and periods T_1 and T_2, periodically coincide at which time $P_i = 1$, as shown by the crosshatched areas. It is assumed that the trains are random (not synchronized) nonharmonic, and $T_1 \neq T_2$. These limitations are necessary to prevent the possibility of noncoincidence. It is clear that if one function $f(A)$, for example, is uncontrolled, as would be the scan of a radar to be intercepted, then the intercept probability can be increased by optimizing function $f(B)$, which could be the passive DF receiver. This optimization would consist of increasing t_2 and decreasing T_2. In a scanning antenna DF receiver, for example, making the antenna

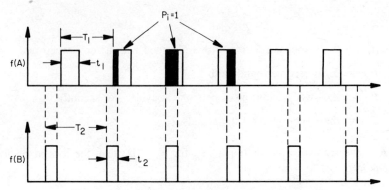

Figure 8-12. Intercept probability for two time window functions.

beamwidth greater would increase t_2; increasing the rotational speed of the antenna would decrease T_2.

The probability of intercept can be stated as follows:

For

$$t_1 \le T_2 - T_1$$

$$P_i = \frac{t_1 + t_2(1 - t_2/2T_1)}{T_2} \quad \text{for } t_2 \le T_1 \tag{8-20}$$

reducing to

$$P_i = \frac{t_1 + T_2/2}{T_2} \quad \text{for } T_1 \le t_2 \tag{8-21}$$

and for

$$T_2 - T_1 \le t_1 \le T_2$$

$$P_i = 1 - \frac{(t_1 - t_2)^2 + (T_2 - t_1)^2}{2T_1T_2} \quad \text{for } t_2 \le T_1$$

reducing to

$$P_i = 1 - \frac{(T_2 - t_1)^2}{2T_1T_2} \quad \text{for } T_1 \le t_2 \tag{8-23}$$

The time Equations (8-20) through (8-23) can represent any scanning function; however, care must be exercised not to compute unreasonable conditions of scan speed, frequency scan as discussed above.

8.5 SUMMARY

The probability of receiving a signal by consideration of the probability-of-detection and the probability-of-intercept has been detailed. An important consideration is that the probabilities themselves may assume certain conditions. The probability of intercept, for example, assumes the independent conditions that the receiver is tuned to the intercept frequency, a beam-on-beam condition exists, and the signal will be received. This also assumes that the probability-of-detection criteria are all met; that is, the threshold above noise has been set for the proper presence of the signal-to-noise ratio consistent with the accuracy required and at the proper system falsealarm rate.

Probabilities multiply, and as a consequence the total probability-of-detection must be computed carefully to be sure that detection takes place within a useful period of time. The scanning antenna, scanning superheterodyne receiver needs high sensitivity, fast antenna rotation and frequency scan, and wide beamwidths. These factors may conflict; for example, high-frequency scan speed may desensitize the receiving process; wide antenna beamwidths may reduce DF accuracy; high-probability omniazimuthal, wide-open frequency receiving may offer higher detection probabilities but at the expense of higher false-alarm rates and possible ambiguities.

In the world of microwave passive DF, the wide-open instantaneous DF driven receiver has become standard for threat warning systems where immediacy of warning is important. These systems may be automatic in their detection, recognition, and response since time is of the essence. In electronic intelligence applications, the scanning or window probability type systems are useful since they can allow time for integration or correlation for processing gain to provide high accuracy. A trade-off is made for use of monopulse techniques on the same basis. Monopulse methods are useful for pulse-by-pulse warning receivers that are DF and/or frequency driven, since monopulse computer sorting provides a relatively fast threat detection decision. For collection systems (ELINT), monopulse methods can be used to improve accuracy and calibrate the environment for preparation of a threat scenario that may be used for the a priori library of the threat warning system.

Our next task will be to study the accuracies that can be expected in various types of systems. These accuracies will reflect back to the method of signal interception and will provide a means to determine the signal-to-noise ratios that will provide a means to determine the signal-to-noise ratios that will be required on an accuracy basis alone. The passive receiver designer will have to consider all factors and perform a systems performance trade-off for a particular design.

REFERENCES

1. Skolnik, M. L., *Introduction to Radar Systems*, New York: McGraw-Hill, 1962.

2. Harp, J. C., "Receiver Performance, What Does Sensitivity Really Mean?" *The International Countermeasures Handbook*, EW Communications, 4th Ed., 1978–1979.

3. Tsui, J. B., *Microwave Receivers with Electronic Warfare Application*, New York: Wiley, 1986, Section 2.4, pp. 20 ff.

4. Papoulus, A., *Probability Random Variables and Stochastic Processes*, 2nd Ed., New York: McGraw-Hill, 1984, pp. 47–48.

5. Skolnik, M. (Ed.), *Radar Handbook*, New York: McGraw-Hill, 1970.

6. Sakaie, Y., "Probability of False Alarm and Threshold of Square-Law Detector-Logarithmic Amplifier for Narrow Band-Limited Gaussian Noise," *IEEE Transactions on Circuits and Systems*, Vol. CAS-33, No. 1, Jan. 1986.

7. Hatcher, B. R., "Intercept Probability and Intercept Time," *Electronic Warfare*, Mar./Apr. 1976.

8. Scheck, P., and A. Behrendt, "Will Intercept Occur." Letters to the Editor, *Electronic Warfare*, Mar./Apr. 1977, p. 44.

9. Self, G. A., "Intercept Time and Its Prediction," *Journal of Electronic Defense*, Aug. 1983, p. 49.

10. Harper, T., "Are Rotating DF Systems Outdated?" *Electronic Warfare*, Mar./Apr. 1974, p. 11.

11. *Spectrum Analyzer Techniques Handbook*, Polarad Electronics Div. of Rohde and Schwartz, Lake Success, NY, 11042, 1962.

12. White, W. D., and I. M. Saffitz, "Compressive Receivers," Topics in Electronics, Vol. 3, 1962, and U.S. Patent 2,954,465.

13. Kinchloe, W. R., "The Measurement of Frequency with Scanning Spectrum Analyzers," Stanford Electronics Lab, Report 557-2, Oct. 1962, Stanford, CA.

14. Tsui, J., "Microwave Receivers and Related Components," National Technical Information Service, Springfield, VA, 1983, PB 84-108711.

Chapter Nine ─────────────

Accuracy of DF Systems

As we have seen, the probability of detection of a receiving system depends upon false-alarm rate, type of receiver technique, the antenna system, and factors established by the target such as range, scan, and time modulation factors. Accuracy depends directly on some of these factors; however, we shall study the problem from the viewpoint of how the receiver system designer can optimize the accuracy with respect to those factors under his control. There is also the question of noise and how it affects accuracy since it is obvious that noise will mask variations representing signal changes, degrading accuracy or requiring higher signal-to-noise ratios that reduce operating sensitivity. The antenna choice and placement, the associated system balance and dynamic range, and the assignment of the error budget are crucial considerations. In any DF system design, there are subtle factors: An aircraft system that must fly a great distance along a line of bearing should have a system that is very accurate and able to be updated as the target is approached. If the system overloads or if the detection process deteriorates as the signal becomes stronger, the mission will be in vain. In a corollary sense, a warning receiver should always give warning, even if the DF measurement is inaccurate due to reception of weak signals, since there will be more time available to improve the accuracy, make decisions, or take action.

Much of the accuracy of a system when mounted in an aircraft or on a ship or other vehicle changes as a function of the environment. The requirement for built-in test facilities (BITE) in most systems tests the operation of the receiver/processor subsystems, but rarely the antennas. Various antenna aperture blockages must be considered in computing the overall effect of the installation. Illumination of the receiver by on-board high-power radars may deteriorate one or more channels of a multichannel receiver, causing a subtle but always present noncatastrophic fault. Periodic calibration and testing must be undertaken to assure that these faults are found and corrected.

This chapter will consider the determination of the accuracy of simple and multibeam receiving systems of the types described previously. Where useful, the mathematical models of the error factors will be presented to aid in the determination of an optimum system choice insofar as the system itself determines the selection.

9.1 ANTENNA ACCURACY FOR MULTIBEAM SYSTEMS

The first discussion of amplitude accuracy will consider the case of DF systems usually categorized as warning or threat identification types (1). These systems are characterized by amplitude comparison of relatively wide-beam antenna patterns with beamwidths of approximately 60–90 degrees. The antennas can be typically mounted at a single point, such as the Honey or a cylindrical array of spirals, or can be conformally dispersed about the host vehicle such as is found in aircraft radar warning applications, where antennas may be wingtip or fuselage mounted. The general idea is to provide a complete 360 degree azimuthal field of view that is distorted as little as possible by the host vehicle. Dispersed antennas such as spirals perform this function well, combining wide RF bandwidth with other desirable attributes such as circular polarization and low cost. For these reasons, this type of antenna will be considered initially. The equations, however, are fully applicable to narrow multibeam arrays whenever a Gaussian approximation for the pattern shapes can be assumed.

9.1.1 The Sine Antenna Pattern Assumption

The ideal antenna to be used in a four-antenna system would exhibit 90 degree 3 dB beamwidths as in Figure 9-1, which shows the E field relationships of a true sinusoidal 90 degree beamwidth pattern. This type of pattern is achievable by Adcock antenna systems and the Honey antenna, where the pattern shapes are relatively independent of frequency over a wide bandwidth, typically an octave.

Assuming that the four antennas are suitably deployed and their outputs detected and logarithmically amplified, it is possible to develop the amplitude monopulse ratio for the subtracted log video signals from Chapter 2, Equation (2-9). Taking the log of the postdetection voltages V_N and V_E induced by the presence of the input signal arriving at ϕ we get the monopulse ratio V_{out}:

$$V_{out} = \log V_E - \log V_N = \log \frac{V_E}{V_N} \qquad (9-1)$$

and substituting for the field patterns
where

$$V_E \qquad = k_E\, B \sin \phi$$

$$V_N \qquad = k_N\, A \cos \phi$$

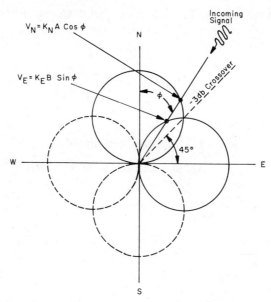

Figure 9-1. Sine field patterns for a four-antenna DF system.

k_E and k_N = the respective channel transfer characteristics

$$V_{\text{out}} = \log \frac{k_E \sin \phi}{k_N \cos \phi} \qquad (9\text{-}2)$$

Assuming $B = A$

or finally

$$V_{\text{out}} = \log \frac{k_E}{k_N} \tan \phi \qquad (9\text{-}3)$$

It may be seen from Equation (9-3) that any common effects such as scan modulation or glint ($B = A$) have been canceled out by the division. This assumes that backlobes of each beam are at a level substantially below the forelobes. This is not always true (see the Honey antenna discussion in Chapter 3), in which case quadrant ambiguity will result. The rule for the subtraction requires that the strongest and next-adjacent strongest antenna be compared, which, when coupled with the backlobe assumption, will assure that the proper pair is being used.

The use of the sine pattern approximation gives rise to the need to balance carefully the gain of the two channels, represented by the voltages k_E and k_N in Equation (9-2). This balance is hard to achieve since the antenna gain of any one lobe varies by only 3 dB over a 45 degree sector, the articulation being only

6.7×10^{-2} dB per degree. This implies a high degree of match ($k_E = k_N$). To illustrate the match required and the general method of development of the error equations, consider

$$\phi^1 = \tan^{-1}\left(\frac{k_E}{k_N}\right)\left(\frac{V_E}{V_N}\right)$$

where

ϕ^1 = the actual angle of arrival

$k_E \neq k_N$, the general case of mismatch

Let

$$X = \frac{k_1}{k_{n+1}}$$

in general where

V_n = the strongest detected lobe

V_{n+1} = the next-adjacent strongest detected lobe

k_1 = the strongest signal

k_{n+1} = the next-adjacent strongest signal

The angular error in degrees is

$$\phi^1 = \tan^{-1} X\left(\frac{V_n}{V_{n+1}}\right)$$

which differs by ϵ, the angular error, such that $\epsilon = (\phi^1 - \phi)$, where ϕ is the true or error-free angle of arrival and $\tan\phi = V_n/V_{n+1}$

$$\epsilon = \tan^{-1} X\left(\frac{k_N V_N}{V_{n+1}}\right) - \phi \tag{9-4}$$

Substituting $\tan\phi$ for V_n/V_{n+1} gives the error

$$\epsilon = \tan^{-1}(X \tan\phi) - \phi \tag{9-5}$$

To determine the worst case error, Equation (9-5) must be differentiated and set equal to zero. Using

$$dx/d \tan^{-1} v = \frac{dv/dx}{1 + v^2}$$

$$\frac{d\epsilon}{d\phi} = \left[\frac{X \sec^2 \phi}{1 + X^2 \tan^2 \phi} - 1 \right]$$

$$= \frac{X \sec^2 \phi - 1 - X^2 \tan^2 \phi}{1 + X^2 \tan^2 \phi} \overset{\Delta}{=} 0$$

and setting the numerator to zero

$$X \sec^2 \phi - 1 - X^2 \tan^2 \phi = 0$$

gives

$$(\tan^2 \phi) X^2 + (\sec^2 \phi) X + 1 = 0$$

Solving for the roots by the quadratic equation gives

$$r_1, r_2 = \frac{-B \pm \sqrt{B^2 - 4AC}}{2A}$$

$$r_1, r_2 = \frac{\sec^2 \phi \pm \sqrt{\sec^4 \phi - 4 \tan^2 \phi}}{2 \tan^2 \phi}$$

But $\tan^2 \phi = \sec^2 \phi - 1$.

Substituting

$$r_1, r_2 = \frac{\sec^2 \phi \pm \sqrt{\sec^4 \phi - 4 \sec^2 \phi + 4}}{2 \tan^2 \phi}$$

$$= \frac{\sec^2 \phi \pm (\sec^2 \phi - 2)}{2 \tan^2 \phi}$$

and using the minus sign since the plus sign gives the trivial solution, $X = 1$. Substituting and solving for X, then ϕ follows

$$X = \frac{2}{2 \tan^2 \phi} = \frac{1}{\tan^2 \phi}$$

or

$$\tan \phi = \pm \frac{1}{\sqrt{X}} \tag{9-6}$$

and

$$\phi = \tan^{-1} \pm \frac{1}{\sqrt{X}} \tag{9-7}$$

Solving (9-6) for $\tan \phi$ and (9-7) for ϕ in Equation (9-5) gives

$$\epsilon = \tan^{-1}(X \tan \phi) - \phi$$

$$= \tan^{-1}(X \tan \phi) - \tan^{-1} \pm \frac{1}{\sqrt{X}}$$

$$= \tan^{-1} X \pm \frac{1}{\sqrt{X}} - \tan^{-1} \pm \frac{1}{\sqrt{X}}$$

$$= \tan^{-1} \pm \sqrt{X} - \tan^{-1} \pm \frac{1}{\sqrt{X}}$$

In the first quadrant

$$\epsilon = \tan^{-1} \sqrt{X} - \tan^{-1} \frac{1}{\sqrt{X}} \tag{9-8}$$

which permits determination of the error in terms of channel unbalance.

Figure (9-2) is a plot of decibel unbalance versus angular error for the sine pattern case of four 90 degree beamwidth antennas. A 1 dB unbalance in gain, for example, will cause a 3.2 degree peak error as shown.

9.1.2   The Gaussian Antenna Pattern Assumption

Now consider a different approximation for the antennas: the use of a Gaussian-shaped pattern. Measurements have shown that spiral antennas, in particular, exhibit beamwidths of this type of the form

$$P = \exp -K\left(\frac{\phi}{\Psi_0}\right)^2 \tag{9-9}$$

where

ϕ = the angle of arrival measured from boresight

Ψ_0 = one-half the 3 dB antenna beamwidth

K = a constant of proportionality

The radiation pattern of these antennas is shown in Figure 9-3. (Note that in the configuration shown, ϕ, the angle of arrival, is measured from the north boresight.)

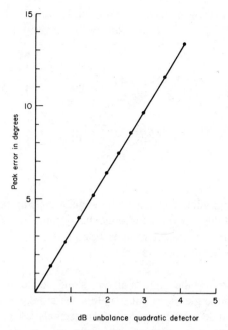

Figure 9-2. DF error versus channel unbalance for a sine pattern assumption (quadratic detector).

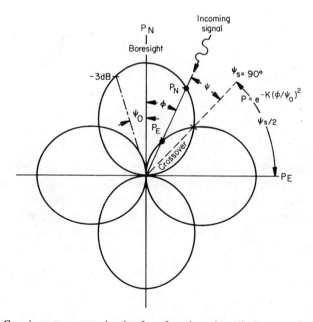

Figure 9-3. Gaussian pattern approximation for a four-channel amplitude monopulse DF system.

The power measured in the east and north antennas from the boresight of each antenna, for example, can be defined from Equation (9-9) as follows:

$$P_E = \exp\left[-K\left(\frac{\Psi_s/2 + \Psi}{\Psi_0}\right)^2\right] \tag{9-10}$$

$$P_N = \exp\left[-K\left(\frac{\Psi_s/2 - \Psi}{\Psi_0}\right)^2\right] \tag{9-11}$$

where

Ψ_s = the squint angle defined here as the angular spacing between the antennas ($360/n = 90$ degrees for a four-antenna system, for example),

Ψ = the angle of arrival measured from the beam crossover point ($\Psi_s/2 - \Psi = \phi$) in the east quadrant, for example

Note: $\Psi_s/2 + \Psi$ means the angle of arrival is outside the crossover of the beams; $\Psi_s/2 - \Psi$ means the angle of arrival is within the crossover of the beams.

Taking the amplitude monopulse ratio R in decibels is accomplished by taking the log of both channels and subtracting as follows:

$$R = 10 \log \frac{P_E}{P_N} \text{ , where } R \text{ is the monopulse ratio}$$

$$= \frac{10\,K \log}{\Psi_0^2}\left[\left(\frac{\Psi_s}{2} + \Psi\right)^2 - \left(\frac{\Psi_s}{2} - \Psi\right)^2\right] \text{ (in decibels)}$$

Solving for Ψ gives

$$\Psi = \frac{\psi_0^2 R}{20 K \Psi_s \log e} \tag{9-12}$$

It is now necessary to obtain the value of the constant K, which can be determined by evaluating Equation (9-9) at the point where the angle of arrival $\Psi = \Psi_s/2 = \Psi_0$ since at this point the response will be -3 dB down. This is the defining level for the antenna beamwidth and by assuming that the crossovers of the beams occur at the 3 dB points.

From (9-9)

$$1/2 = e^{-k\left(\frac{\Psi}{\Psi_0}\right)^2}$$

Substituting

$$\Psi = \Psi_0$$

$$1/2 = e^{-k}$$

$$K = 0.69$$

replacing k in Equation (9-12), and simplifying gives

$$\Psi = \frac{\Psi_0^2}{6\Psi_s} R \tag{9-13}$$

Equation (9-13) shows that the angle of arrival Ψ measured from the crossover angle $\Psi_s/2$ is a function of the square of the antenna beamwidth and is inversely proportional to squint angle or spacing; it is constant with azimuth. If Equation (9-13) is solved for R,

$$R = \frac{6\Psi_s \Psi}{\Psi_0^2}$$

and if R is differentiated with respect to Ψ to determine the error slope, which is defined as the rate of change of the ratio as a function of the angle of arrival (Ψ is directly related to ϕ by definition above), then

$$\frac{dR}{d\Psi} = \frac{6\Psi_s}{\Psi_0^2} = C \text{ (a constant)}$$

The error slope is determined *only* by the squint (Ψ_s) and the (Ψ_0) antenna beamwidths for Gaussian-form expressions of antenna patterns. This means that ϕ and Ψ are only related to selected constants. If R is measured as the ratio of the strongest to next adjacent strongest signal, it can yield these angles by comparison to a table of values independent of the angle of arrival. This is true for all antenna patterns that can be expressed in exponential form.

The above conclusion permits the use of one look-up table for a DF system using antennas that exhibit essentially constant beamwidths, such as the spiral, as long as the pattern can be reasonably expressed in the Gaussian form. For passive detection, any Gaussian antenna that is essentially frequency stable in beamwidth, in theory, eliminates the need for a frequency word to correct the DF measurement. This property makes the broadband spiral useful for radar warning applications over wide frequency ranges (10:1). When the beamwidths of the spirals do vary (at the lower band edge), band division filters, such as multiplexers as discussed in Chapter 4, may be used with associated look-up tables assigned to each band to improve accuracy due to frequency dispersive beamwidths.

Channel Unbalance Accuracy. The unbalance error or difference in gain in decibels between the two operating channels can be determined by calculating $d\Psi/dR$ from Equation (9-13). This is the partial derivative of Equation (9-13). It gives the angle of arrival Ψ as a function of three variables: R, the log ratio, Ψ_0 the one-half beamwidth, and Ψ_s the fixed squint angle:

$$\Psi = f(R, \Psi_0, \Psi_s) = \frac{\Psi_0^2 R}{6\Psi_s} \tag{9-14}$$

Differentiating for ϵ_R, the unbalance error, gives

$$\frac{d\Psi}{dR} = \frac{\Psi_0^2}{6\Psi_s} = \epsilon_R \tag{9-15}$$

The results are plotted in Figure 9-4 for a four-antenna ($\Psi_s = 90$ degrees) DF system and in Figure 9-5 for a six-antenna ($\Psi_s = 60$ degrees) DF system. From these curves at a given antenna beamwidth Figure 9-6 presents the results for a 1 dB unbalance, a common number used in the error budget for systems of this type. The curves can be used two ways: It is possible to determine the amount of error

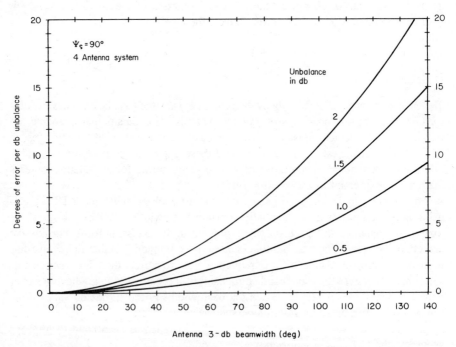

Figure 9-4. DF error (e_r) as a function of decibel gain unbalance in a four-antenna ($\Psi_s = 90$ degree) DF system.

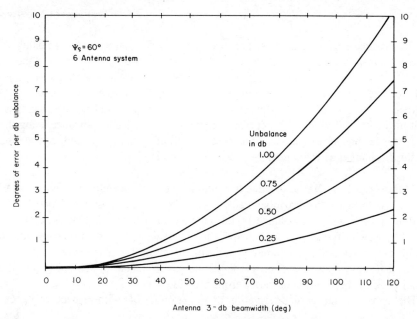

Figure 9-5. DF error (e_r) as a function of decibel gain unbalance in a six-antenna (Ψ_s = 60 degree) DF system.

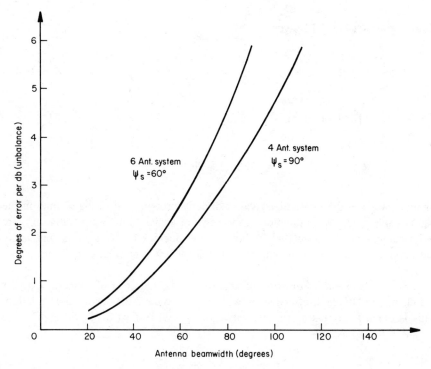

Figure 9-6. Antenna pointing error as a function of antenna beamwidth for a four- and six-antenna system for a 1 dB unbalance.

due to unbalance in the channels. For example, in a four-antenna system using 70 degree beamwidths, a 1 dB unbalance would cause 2.29 degrees of error if the 70 degree antenna increased in beamwidth to 100 degrees (at the low-frequency band end, for example). The error in degrees for a 1 dB unbalance would equal 4.63 degrees, making the use of a different look-up table, chosen by a low-band bandpass multiplex filter, advisable. The curves assume a positive 1 dB unbalance using the lower valued channel as reference; if a negative 1 dB unbalance were used, considering the upper channel as reference, the error would be negligibly different.

Squint Accuracy. Variations due to pointing accuracy or squint are important to consider since squint for broadband antennas will vary with frequency. Also, in the case of spirals, with the balance of the antenna feed network (balun). Equation (9-13) can be differentiated with respect to Ψ_s to determine ϵ_s, the error due to squint, the actual variation of antenna pointing direction. For planar spiral antennas, this is the angle between a line perpendicular to the plane of the spiral and the actual peak of the beam as it varies from this line.

The error ϵ_s is the partial derivation of Equation (9-14) with respect to Ψ_s, or

$$\frac{\partial \Psi}{\partial \Psi_s} = -\frac{R \Psi_0^2}{6 \Psi_s^2} = \epsilon_s$$

Substituting R from 9-13 gives

$$\frac{\partial \Psi}{\partial \Psi_s} = -\frac{\Psi}{\Psi_s} \tag{9-16}$$

The amount of the variation is

$$\epsilon_s = -\left(\frac{\Psi}{\Psi_s}\right) \Delta \Psi_s \tag{9-17}$$

or the change in the angle of arrival is directly proportional to the change in the squint angle modified by the ratio of the angle of arrival to the squint angle. Figure 9-7 is a plot of Equation (9-17) showing the degrees of error per squint angle variation as a function of the angle-of-arrival for different values of squint.

Antenna Beamwidth Variation in Accuracy. The last error of concern, ϵ_{Ψ_0}, is the change in DF angle-of-arrival Ψ with respect to the 3 dB beamwidth variation. This is obtained by taking the derivative of Ψ with respect to Ψ_0 or again using Equation (9-13) as follows:

$$\frac{\partial \Psi}{\partial \Psi_0} = \frac{2 \Psi_0 R}{6 \Psi_s} \tag{9-18}$$

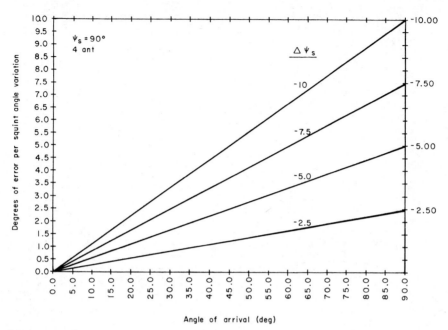

Figure 9-7. DF error per degree of squint variation as a function of angle of arrival for a 90 degree beamwidth Gaussian pattern spiral.

$$\frac{\partial \Psi}{\partial \Psi_0} = \frac{2 \Psi_0 R}{6 \Psi_s} \tag{9-18}$$

Substituting for R from (9-13) gives

$$R = \frac{6 \Psi \Psi_s}{\Psi_0^2} \tag{9-19}$$

$$\epsilon_{\Psi_0} = \frac{2 \Psi}{\Psi_0} \tag{9-20}$$

Plotting the error gives Figure 9-8, which shows the beamwidth variation in a four-antenna 90 degree squint angle system.

The total RMS error for a DF system, assuming no correlation, is

$$\Delta \Psi_{RMS} = \sqrt{(\epsilon_R \Delta R)^2 + (\epsilon_{\Psi_0} \Delta \Psi_0)^2 + (\epsilon_{\Psi_s} \Delta \Psi_s)^2}$$

It is often useful to determine the boresight-to-crossover loss in adjacent patterns in a DF system since this is the value of change from the boresight that must be encoded in the look-up table. At the point of crossover, the next ratio may be formed; e.g.,

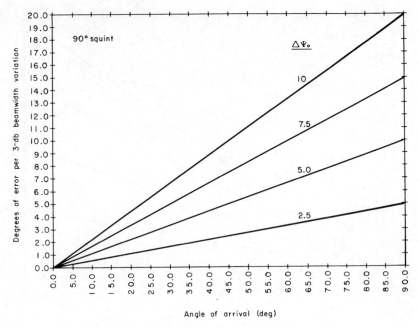

Figure 9-8. DF error due to variation of 3 dB beamwidth.

$$R = \frac{P_N}{P_E}\Bigg]_{\phi=0}^{\phi=45} \quad ; R = \frac{P_E}{P_N}\Bigg]_{\phi=45}^{\phi=90}$$

It is, however, desirable to control antenna characteristics well into the region of the next beam since errors are computed by the straight line logarithmic power ratio. The crossover loss can be determined from Equation (9-9), or

$$P(\text{dB}) = \exp\left[-k\left(\frac{\phi}{\Psi_0}\right)^2\right] \qquad (9\text{-}21)$$

In decibels,

$$P(\text{dB}) = 10\log\exp\left[-k\left(\frac{\phi}{\Psi_0}\right)^2\right] \qquad (9\text{-}22)$$

Letting r^1 equal the ratio

$$r^1 = -3\left(\frac{\phi}{\Psi}\right)^2 \qquad \text{(in decibels)}$$

To check test the case at 45 degrees for a 90 degree beamwidth pattern,

$$r^1 = 3 \left(\frac{45}{90/2}\right)^2 = 3 \text{ dB}$$

which is correct since a 90 degree beamwidth is -3 dB down from the beam peak at the half power point. For a 45 degree crossover in a four-antenna 90 degree squinted system, the maximum variation ($\phi = \Psi_s = 90$ degrees; $\Psi_0 = 45$ degrees) is

$$R^1 = -3 \left(\frac{90}{45}\right)^2$$

$$R^1 = -12 \text{ dB} \qquad \text{in RF ratio}$$

This means that the total variation from one cardinal axis to another ($\phi = 90$ degrees) between two 90 degree beamwidth patterns is 12 dB, which is the decibel pattern variation that should be entered in the look-up conversion table for a receiver of this type.

Figure 9-9 is a plot of the crossover sensitivity loss for a four-beam and six-beam system for antennas squinted at 90 and 60 degrees, respectively. If the squint

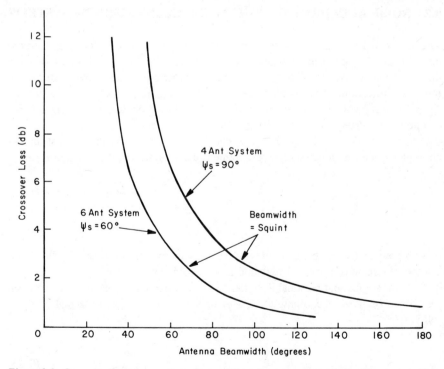

Figure 9-9. Crossover loss versus antenna beamwidth at the crossover points for four- and six-beam antenna systems.

angle is equal to the antenna beamwidth, the maximum mainlobe to crossover loss will always be four times the crossover loss in decibels. If the values differ, the loss must be computed. For example, a typical spiral has a beamwidth of 70 degrees and in a four-channel warning system, the crossover loss is actually

$$(r^1)_{45} \text{ degrees } = -3\left(\frac{45}{70/2}\right)^2 = -4.96 \text{ dB}$$

and the maximum variation is

$$(r^1)_{45} \text{ degrees } = -3\left(\frac{90}{70/2}\right)^2 = -19.84 \text{ dB}$$

For only a small increase in the 45 degree crossover loss, the articulation of the antenna is improved, relieving the burden of obtaining balance in channels, axial ratio balance, and so forth. Familiarity with these trade-offs will facilitate assignment of error budgets and will permit choice of beamwidths and antenna types. A multibeam planar or circular array can be analyzed using the above procedure assuming Gaussian patterns for the lobes.

9.2 NOISE ACCURACY OF AMPLITUDE COMPARISON DF SYSTEMS

In Chapter 8, Section 8.1, we established that the probability of detecting a *signal* depends upon where the voltage threshold is set. We have established that the threshold can be set to obtain a given false alarm rate. This may often be the minimum threshold. The operating setting, however, may depend upon a desired signal quality, not necessarily the FAR or probability of detecting a signal. For example, in a true monopulse analog DF system, it may be required to obtain a 0.5 degree resolution using 70 degree beamwidth antennas requiring that an unambiguous determinable change be obtained. This would require a signal-to-noise ratio at video of at least

$$10 \log \frac{70}{0.5} = 21 \text{ dB}$$

It is obvious that with no signal integration (the true monopulse case), 21 dB would impose severe sensitivity and dynamic range problems.

If it is assumed that reasonable signal-to-noise ratios will be used, then Equation (9-13) can be used to compute the DF error due to uncorrelated thermal noise as follows:

$$\Psi = \frac{\Psi_0^2}{6\Psi_s} R \text{ (dB)} = \frac{\Psi_0^2}{6\Psi_s} 10 (\log S_1 - \log S_2) \tag{9-23}$$

Replacing R by 10 ($\log S_1$ − $\log S_2$), where S_1 and S_2 are signal peak powers, since the powers are proportional to the square of the voltages, gives

$$\Psi = \frac{20\Psi_0{}^2}{6\Psi_s} [\log(V_1) - \log(V_2)]$$

where V_1 and V_2 are the signal voltages (after detection).

The change in Ψ with respect to V_1 and V_2 can be obtained by taking the partial derivatives as follows:

$$\frac{\partial\Psi}{\partial V_1} = \frac{20\Psi_0{}^2}{6\Psi_s}\left(\log e\right)\left(\frac{1}{V_1}\right) \tag{9-24}$$

$$\frac{\partial\Psi}{\partial V_2} = \frac{20\Psi_0{}^2}{6\Psi_s}\left(\log e\right)\left(\frac{1}{V_2}\right) \tag{9-25}$$

Figure 9-10 shows the peak output voltage change for an input noise voltage ($n\sqrt{2} = \Delta V$). This is the approximate case of a small input increment, approximating the differential (limiting) change. The n_1 and n_2 values are RMS noise voltage. It

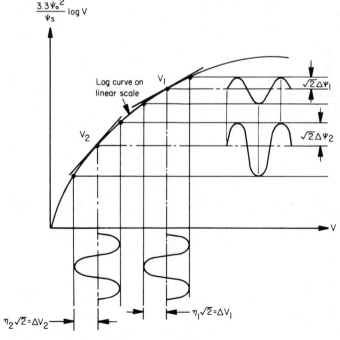

Figure 9-10. Relationship between the change in the rear output noise voltage for the two RMS input voltages ΔV_1 and ΔV_2 using the increment Δ to approximate the derivative.

may be seen that since the transfer characteristics of the amplifier (which is logarithmic) is plotted linearly, there is a smaller output noise voltage at a higher input level (V_2 is noisier than V_1). We are therefore considering a fairly good signal-to-noise ratio, which would be needed to satisfy the FAR and P_D criteria X_0 to permit the use of a Gaussian noise distribution. These curves and the calculations are accurate to the degree that this assumption holds. Solving Equations (9-24) and (9-25) for Ψ in terms of RMS values gives

$$\Delta\Psi_1 = \frac{20\Psi_0^2 \log e}{6\Psi_s} \frac{1}{V_1} \frac{\Delta V_1}{\sqrt{2}} \tag{9-26}$$

and

$$\Delta\Psi_2 = \frac{20\Psi_0^2 \log e}{6\Psi_s} \frac{1}{V_2} \frac{\Delta V_2}{\sqrt{2}} \tag{9-27}$$

For the RMS $\Delta\Psi_1$ and $\Delta\Psi_2$ values, Equations (9-26) and (9-27) must be summed as the square root of the sum of the squares (Gaussian assumption):

$$(\Delta\Psi)_{RMS} = \sqrt{\Delta\Psi_1^2 + \Delta\Psi_2^2} \tag{9-28}$$

Simplifying and substituting Equations (9-26) and (9-27) into Equation (9-28) yields

$$(\Delta\Psi)_{RMS} = \frac{1.446\Psi_0^2}{\Psi_s} \sqrt{\frac{\Delta V_1^2}{2V_1^2} + \frac{\Delta V_2^2}{2V_2^2}} \tag{9-29}$$

But $\Delta V_1^2/2$ and $\Delta V_2^2/2$ can be recognized as the video noise powers N_1 and N_2, and V_1^2 and V_2^2 are the signal powers in channels 1 and 2 used to form the monopulse ratio R. Then, substituting, we get

$$(\Delta\Psi)_{RMS} = \frac{1.446 \ \Psi_0^2}{\Psi_s} \sqrt{\frac{N_1}{S_1} + \frac{N_2}{S_2}} \tag{9-30}$$

Assuming $N_1 = N_2 = N$, where N_1 and N_2 are uncorrelated noise, gives

$$(\Delta\Psi)_{RMS} = \frac{1.446\Psi_0^2}{\Psi_s} \sqrt{\frac{N}{S_1} \left(1 + \frac{S_1}{S_2}\right)} \tag{9-31}$$

Relating the noise error to the ratio of power. But S_2/S_1 is the *monopulse* power ratio from Equation (9-23) or $S_2 = g(\Psi)S_1$.

The equation

$$g(\Psi) = 10^{-\left(\frac{\Psi\Psi_s}{1.66\Psi_0^2}\right)}$$

can be derived by manipulating Equation (9-23). Finally, the DF error due to uncorrelated thermal noise is obtained by substitution

$$\Delta \Psi = \frac{1.446 \Psi_0^2}{\psi_s} \sqrt{\frac{1}{\text{SNR}} \left(1 + \frac{1}{g(\Psi)} \right)} \qquad (9\text{-}32)$$

where SNR is the video signal-to-noise ratio

The values for $\Psi = 45$ degrees and $\Psi_s = 90$ degrees can be substituted for the four-beam system giving the error $\Delta \Psi$ as a function of video signal-to-noise ratio and angle of arrival Ψ as

$$\Delta \Psi = 32.5 \sqrt{\frac{1}{\text{SNR}} (1 + 10^{.026 \ \Psi})} \qquad (9\text{-}33)$$

which is plotted in Figure 9-11 in two scales and for signal-to-noise ratios of 10–20 dB in the lower curve and 25–35 dB in the upper curve.

It is interesting to note that the average error is approximately equal to that obtained from a rule-of-thumb method. For example, a 20 dB signal-to-noise ratio

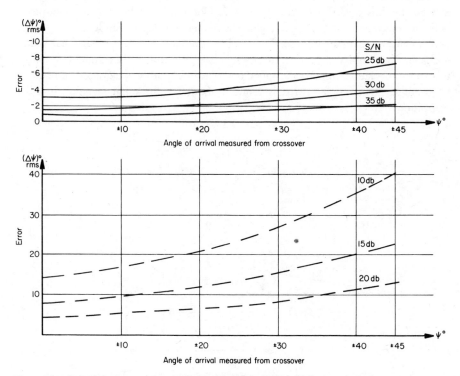

Figure 9-11. DF error versus angle of arrival Ψ measured from crossover for various video signal-to-noise ratios.

is 10:1; therefore, in a 90 degree beamwidth system there will be a 90 degree/ 10 = 9 degree average error. From the curves it may be determined that the average DF error for 20 dB signal-to-noise ratio, of all of the errors, plus and minus, is actually 8.25 degrees at the average angle of approval Ψ of 24 degrees.

The curves permit the determination of the RMS error at *each* angle of arrival, remembering that Ψ is measured from crossover compared to the more familiar ϕ measured from boresight. These curves may be redrawn for other combinations of Ψ_0 and Ψ_s. The curves are also useful for linear multibeam array patterns that can be approximated by a Gaussian curve.

9.3 NOISE ACCURACY IN PHASE MEASUREMENT DF SYSTEMS

The accuracy of phase measurement systems is directly dependent upon the signal-to-noise ratio since any phase noise will create an ambiguity in proportion to the magnitude of phase shift it causes. If we consider that phase between two sine waves can be measured by measuring the difference in time at some point at the same place on each sine wave, say the zero crossing axis, then any ambiguity in zero crossing of one sine wave to another is an error. If both sine waves have phase noise, defined here as zero crossing perturbation, then the ambiguity will double.

Random band-limited noise can be represented as a vector with two components that can be drawn as shown in Figure 9-12. One noise component, n_i, is in phase with the amplitude of the signal, while a second 90 degree shifted noise vector component, n_q, is in quadrature. The in-phase component adds to the signal vector $E \cos \omega t$, varying its amplitude but not its phase. In limiter/discriminator systems, this variation can be limited or stripped out. The effect of the quadrature component is to add a phase angle γ that is random about E_r, the input signal, where by inspection

$$\gamma \cong \tan^{-1} \frac{n_q}{E + n_i} \qquad (9\text{-}34)$$

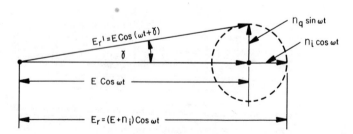

Figure 9-12. Vector relationships for the effect of noise on phase.

For large signal-to-noise ratios $E > n_i$, $E > n_q$,

$$E_r \cong E_r{}'$$

and

$$\tan \gamma = \gamma \qquad \text{(approximately)}$$

then

$$\gamma \cong \frac{n_q}{E} \tag{9-35}$$

The total signal power is proportional to the square of E_r:

$$S = E_r{}^2$$
$$n_n{}^2 = n_I{}^2 + n_q{}^2 \qquad \text{(the noise power)} \tag{9-36}$$

where

$$n_n{}^2 = \text{the RMS noise power}$$
$$n_i{}^2 = \text{proportional to the } I \text{ channel RMS noise power}$$

and

$$n_q{}^2 = \text{proportional to the } Q \text{ channel RMS noise power}$$

Since the noise is symmetrically distributed about the carrier ω_c,

$$n_n{}^2 = n_q{}^2$$

Since the total noise power

$$n_n{}^2 = n_I{}^2 + n_q{}^2 = 2n_q{}^2 \qquad \text{(by substitution)}$$

solving for $n_q{}^2$ gives

$$n_q{}^2 = \frac{n_n{}^2}{2} \tag{9-37}$$

As stated above, the only component of the noise creating the phase angle γ is the quadrature noise forming the ratio for γ as follows:

$$\gamma = \frac{n_q}{E_r} = \sqrt{\frac{1}{[(E_r)^2/(n_q)^2]}} = \sqrt{\frac{1}{[2E_r{}^2/n_n{}^2]}} \qquad (9\text{-}38)$$

Letting

$$S = \text{total signal power; } E_r = \sqrt{s}$$

$$N = \text{total noise power; } N = n_n^2$$

then

$$\gamma = \frac{1}{\sqrt{2} \; S/N} \text{ in radians } (180/\pi \text{ in degrees}) \qquad (9\text{-}39)$$

If the phase between two uncorrelated noise signals is measured, then the total differential RMS variance $\gamma_{T(\text{RMS})}$ is

$$\gamma_{T(\text{RMS})} = \sqrt{N_1{}^2 + N_2{}^2}$$

$$= \sqrt{\frac{1}{2 \; S/N} + \frac{1}{2 \; S/N}}$$

Assuming the signal-to-noise ratio (S/N) values are equal,

$$\gamma_T = \frac{1}{\sqrt{S/N}} \qquad \text{in radians } (180/\pi \text{ in degrees}) \qquad (9\text{-}40)$$

It is useful to reference the above phase noise perturbations to angles measured from boresight as in the common case. Figure 9-13 is a two-element interferometer consisting of antennas A_1 and A_2 separated by D. The angle the plane wave makes with the boresight is ϕ.

If

$$V_A = 1 \underline{|\theta_A}$$

and

$$V_B = 1 \underline{|\theta_B}$$

where the phases θ_A and θ_B are relative to boresight, then

$$\theta_A = \frac{\pi D}{\lambda} \sin \phi = +\theta \qquad (9\text{-}41)$$

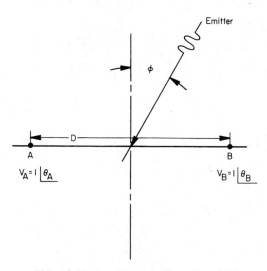

Figure 9-13. Interferometer phase noise model.

$$\theta_B = \frac{\pi D}{\lambda} \sin \phi = -\theta \qquad (9\text{-}42)$$

The effects of this noise variation in the measurement of the angle of arrival ϕ of a two-antenna interferometer can be calculated differentiating Equations (9-41) and (9-42) with respect to ϕ:

$$\frac{d\theta_A}{d\phi_A} = \frac{\pi D}{\lambda} \cos \phi; \quad d\phi_A = \left(\frac{\lambda}{\pi D \cos \phi}\right) d\theta_A \qquad (9\text{-}43)$$

$$\frac{d\theta_B}{d\phi_B} = \frac{-\pi D}{\lambda} \cos \phi; \quad d\phi_B = \left(\frac{\lambda}{\pi D \cos \phi}\right) d\theta_B \qquad (9\text{-}44)$$

The DF error is

$$\Delta\phi = \sqrt{\Delta\phi_A{}^2 + \Delta\phi_B{}^2} \qquad (9\text{-}45)$$

Letting $d\phi = \Delta\phi$ and substituting from Equations (9-43) and (9-44) gives

$$\Delta\phi = \frac{\lambda}{\pi D \cos \phi} \sqrt{\Delta\theta_A{}^2 + \Delta\theta_B{}^2} \qquad (9\text{-}46)$$

where $\Delta\theta_A = \Delta\theta_B = \gamma$, the phase error in Equation (9-46). Then

$$\Delta\phi = \frac{\lambda}{\pi D \cos \phi \sqrt{S/N}} \text{ radians} \qquad (9\text{-}47)$$

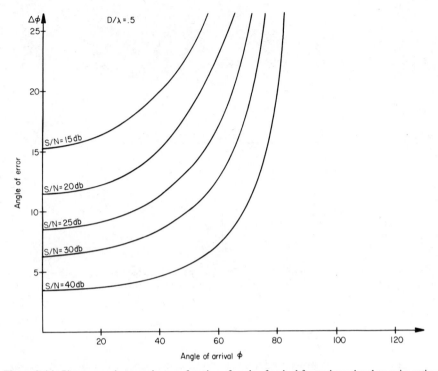

Figure 9-14. Phase error due to noise as a function of angle of arrival for various signal-to-noise ratios for $d/\lambda = 0.5$.

Figure 9-14 is a plot of the case of $D/\lambda = \frac{1}{2}$ showing the phase error in the monopulse ratio as a function of input signal-to-noise ratio and angle of arrival. The effect of this noise voltage is to fill in the null in a $(\Sigma-\Delta)$ monopulse system and to shift the zero crossings or phase differences as a function of $1/\sqrt{S/N}$. As mentioned before in the circular phase interferometer, the mode summation improves the signal-to-noise ratio in each channel by n, where n is the multiplier of ϕ, the angle-of-arrival. For a given signal-to-noise ratio, the least significant bit will have the least error and the most significant bit the greatest error, leading to the conclusion that a combination of amplitude and phase monopulse techniques is desirable; the amplitude monopulse system determines the most significant bit; the phase encoder, the least.

9.4   ACCURACY AS A FUNCTION OF INSTALLATION

It is important as part of the general discussion of accuracy of passive DF systems to consider the location or sighting of the antennas as a fundamental contributor to overall accuracy. It is unreasonable to expect that antenna pattern measurements made in an anechoic chamber or under ideal conditions will be duplicated in a field installation. In some instances, the management of the deterioration of DF performance on the host vehicle can be a major factor in the choice of the appropriate DF system, more so perhaps, than the expected performance from theoretical considerations considered alone. Configuration of direction-finding systems also depends upon whether or not the purpose is long-range angle-of-arrival determination or immediate DF information for warning purposes and can be further classified on the basis of intercept probability, angular field of view, response time, and accuracy.

Rotating DF antenna systems make use of more sophisticated receivers since there are fewer antennas, thereby reducing the need for multiple channels. The higher sensitivity, thus available, may permit reception of side and back lobes of the signal to be detected, trading receiver cost for complexity. Although a higher sensitivity receiver would imply that more signals could be received, the sensitivity usually is the by-product of a narrower RF bandwidth, which as we have seen may require frequency scanning. If time is not a problem, the technique can be used to advantage to unburden the processing system, since the amount of data made available is limited by the scan.

Fixed antennas may be located at a single point or geographically dispersed about the host vehicle. In aircraft-mounted radar warning receivers, low gain (0 dBi) wide band (2–18 GHz) spiral antennas are dispersed about an aircraft effectively enclosing the vehicle to minimize effects of the unsymmetrical ground plane formed by the aircraft wings, fuselage, and so on. Very high sensitivity is not required in warning applications, but since 360 degrees of azimuth coverage over a wide-frequency range is provided, signal densities can become high enough to present serious digital processing difficulties. Direction-finding systems are also greatly dependent upon the configuration of the ground plane structure around the antennas, reflections from surrounding surfaces, and interference from structures or shadowing in the field of view. The most desirable locations for a fixed-location system, for example, might be at the top of a ship's mast or for hemispherical coverage on an aircraft at the lowest point on the fuselage. These locations are rarely available.

Figure 9-15, a plan view diagram of a typical ship, illustrates the type of problem often encountered in naval installations. It would appear that the top of the mast system would be ideal; however, the length-to-width ratio of the ship's ground plane creates pattern distortion and can still cause a problem. Other considerations are shadowing effects due to other masts, radar antennas, and obstructions.

Reflections can occur due to multipath transmissions in smooth sea states, which give rise to inaccurate pulse width and DF measurements, both of which can cause processing problems unless precision frequency of arrival is also used as a primary sorting tool. One solution to the blockage problem is an antenna system that is sectorized into a port and starboard array, which can be mounted on an open face of the superstructure to provide a clear angle of view. In Figure 9-15, 150 degree sectors of high-accuracy coverage is shown for each side, leaving only two 30 degree sectors fore and aft to be covered with lower accuracy 30 degree horn antennas. The sector antennas can be lens-fed arrays or a set of parallel horns. These types of systems offer many advantages:

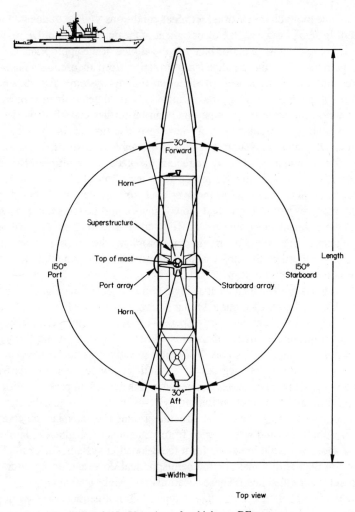

Top view

Figure 9-15. Plan view of a shipborne DF system.

1. High DF accuracy is available port and starboard, where the ship exhibits maximum cross-sectional area to a radar.
2. The reduced DF accuracy fore and aft occurs at minimum ship's radar cross section.
3. With good design it is often possible to eliminate side- and backlobe inhibition channels, since the ship's superstructure isolates each wide field-of-view array.

In the use of dispersed antennas, the ground plane effects of the ship are minimized. The dispersed antenna system is generally more expensive; however, since many receiver channels are required, a more accurate and site-independent system can be configured.

Although sectorized DF systems are perhaps the most accurate, single-point DF systems work well on smaller ships and are very popular where a true top-of-the-mast location can be found. The Anaren fixed Butler-fed circular array in a radome, as described previously, when mounted on a frigate in the position shown in Figure 9-16 (2) was able to attain an RMS accuracy of 1.7 degree RMS in the 8–16 GHz band over a wide variation of angle-of-arrival with respect to the ship. The photograph shows what must be considered a nearly ideal location, with minimum aperture blockage, accounting in part for this excellent performance, despite the abovementioned ground plane effects.

In aircraft installations, it is desirable to obtain a field-of-view that extends above and below the horizon to detect the presence of air-to-air and ground-to-air threats (*H*-plane coverage) as well as provide a 360 degree omniazimuthal DF capability. The use of four or six geographically dispersed spiral antennas is a common solution for this requirement. A single point mounted antenna system may also be used for high accuracy applications, usually at the price of spatial coverage. Figure 9-17 shows a subsonic aircraft outlining antenna location possibilities. Location A shows a belly-mounted radome enclosed rotary antenna. Location B depicts a wing-tip mount where two or three spirals can be oriented to provide 180 degree coverage for each side. Locations C and D show how forward and rear quadrants can be covered by mounting two spirals in a radome (location C) and two antennas in the tail structure (location D) for complete 360 degree azimuthal coverage. In the subsonic aircraft shown, fiberglass radomes are often used to provide protection.

Figure 9-16. Top of mast location for sea trials of Anaren circular array. Courtesy of Anaren Microwave Corp.

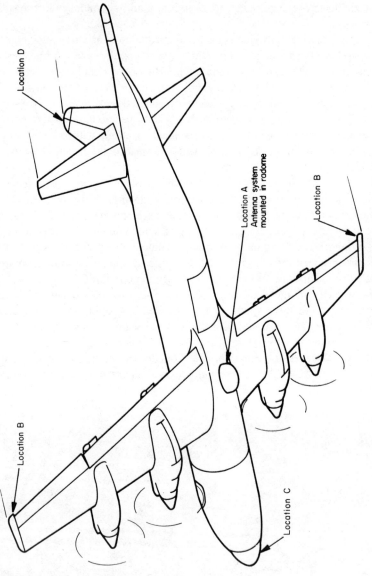

Location D

Location A
Antenna system
mounted in radome

Location B

Location B

Location C

Figure 9-17. Typical DF antenna locations on aircraft.

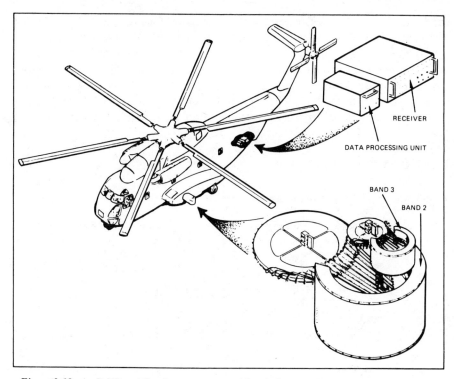

Figure 9-18. An R-KR multiband antenna mounted in a helicopter. Courtesy of Raytheon Corp.

subsonic aircraft shown, fiberglass radomes are often used to provide protection. In supersonic aircraft, flush-mounted antennas may be dispersed around the aircraft in a similar manner; however, temperature resistant radomes must be used. The proximity of aircraft stores such as pods, armament, and other equipment may cause shadowing effects that will definitely affect accuracy. Figure 9-18 shows an R-KR multiband antenna mounted in a helicopter location that provides 180 degree sector coverage. In this case, high accuracy is obtained by designing the overall helicopter's installation as a platform for the DF function. Radar warning antennas in helicopter installations follow the general rules for subsonic aircraft.

DF accuracy of airborne systems often suffer from antenna depolarization effects, especially for flush-mounted antennas. Signals that are horizontally polarized see an equipotential surface along the conducting surface skin of the aircraft, which cancels the horizontal field component leaving only the vertical. Under these conditions, a circularly polarized antenna will tend to exhibit linear or elliptical polarization along the fuselage line, causing a null along the aircraft and "grating" or multiple-null lobes. For this reason, it is desirable to disperse spiral antennas at wing and tail tips in a space-quadrature orientation to provide a field of view perpendicular to the plane of the spirals. Perturbations in patterns can be compensated for by anechoic chamber calibrations of aircraft or by site calibrations for ships. Corrections of DF measurement as a function of received azimuth or as a function of frequency can be applied to the DF measurement computation for high-precision or, in the case of interferometer systems, ambiguity resolution. Digital correction is becoming a far more economical solution to improve accuracy than trying to improve the antennas or vehicle mounting locations. This technique is finding favor in those installations using extensive digital processing where calibration can be accomplished. Other correction methods include the measurement of the elevation angle (θ) as well as the azimuth (ϕ) to provide corrections for aircraft maneuvering.

REFERENCES

1. Lipsky, E., "Understanding DF Accuracy of Radar Warning Receivers," *Electronic Warfare*, *Defense Electronics*, Apr. 1978.
2. "DESM Receiver Sea Trials," Anaren Microwave, Inc., E. Syracuse, NY 13057, 1985.

Chapter Ten

Signal Processing
and Display Methods

The various methods and tools of the art of microwave direction finding have been presented with many details. Like all complex designs, there is often a single forcing function that governs. In some cases, it is performance; in others, size; in most, it is cost. To be complete, it is also necessary to be familiar with how the outputs of the DF system will be used in conjunction with other subsystems. Many of the techniques, methods, and devices discussed in this book are abstract. This chapter will examine the methodology of significant hardware aspects of passive direction finding from a practical viewpoint; that is, the analog processes of detection and amplification over wide dynamic DF range, associated signal processing, and the popular forms of radar warning and electronic intelligence DF display.

10.1 LOGARITHMIC AMPLIFIERS

The logarithmic amplifier finds wide use in passive direction finding systems and can be a major factor in performance/cost trade-offs. These amplifiers are used to perform two essential functions. The first is the compression of a wide input dynamic range to a manageable output dynamic range; the second is to provide a convenient mathematical method to form the monopulse ratio for normalization. This is done by subtraction of the logarithmic outputs that form the logarithmic fractional ratio. In passive direction finding, the wide variation of pulse characteristics and the need to cover large instantaneous RF bandwidths have channeled the evolution of the logarithmic amplifier into specific configurations.

There are four basic types of logarithmic amplifiers: the logarithmic video amplifier, the compressive logarithmic video amplifier, the logarithmic RF amplifier, and the successive detection logarithmic IF amplifier. Each type has certain reasons for use with DF systems. From an overview, the logarithmic video amplifier is an analog amplifier with a logarithmic input/output transfer characteristic designed to complement the square-law RF-to-video conversion characteristics of a detector. In combination with a single detector, this amplifier is known as a detector logarithmic video amplifier (DLVA) and is characterized by wide RF and video bandwidth and excellent dynamic range. One variation of this, the second type, is the log–log amplifier that provides even more dynamic range. Since all phase information contained in the carrier is lost in the detection process, the DLVA is basically only suitable for analog/class II processing. The third type, the RF logarithmic amplifier, is an RF configuration that operates as a true logarithmic amplifier at the RF or carrier frequency. Since this amplifier does not detect, all RF phase information can be preserved permitting use in phase recognition applications analogous to the class I (IAGC) type monopulse processor. The fourth type of amplifier utilizes successive detection of the RF signal and summation of the resultant video. It develops a logarithmic characteristic from the saturating (or cutoff) gain effects of each of a series of stages. Since a stage can only contribute a fixed output past a certain dynamic range, an arithmetic sum of these outputs occurs for a geometric input variation, resulting in the desired logarithmic response. The successive detection amplifier finds application as an IF amplifier in most cases and, as described elsewhere in this book, is popular for radar and narrow bandwidth scanning receivers. Here again only amplitude information is preserved.

10.1.1 The Detector Logarithmic Video Amplifier

The most common type of logarithmic amplifier is the detector logarithmic video amplifier, shown in Figure 10-1. Most DLVAs consist of a wide RF bandwidth detector preceded by an equally wideband antenna and protective limiter, and sometimes an RF preamplifier. The antenna is usually a spiral, which can cover 2–18 GHz providing a 16 GHz instantaneous bandwidth. This RF bandwidth can be multiplexed into sub-bands for octave band coverage (2–4, 4–8, 8–16 GHz), as shown in Chapter 4, Figure 4-5, in various configurations. The DLVA offers the advantages of simplicity, wide bandwidth, and relatively good tracking between units. It is ideally suited to class II monopulse (subtracted logarithmic amplifier) applications where it can advantageously recover the information bandwidth of the radar video signal if a sufficiently large B_v is used. The disadvantages are that it is not sensitive, unless preceded by an RF amplifier, and has a somewhat limited dynamic range. This may be overcome by splitting the input signal, adding gain to one channel, and summing both outputs. Another problem of the DLVA may be DC drift. If CW or high duty cycle environments are to be received, DC coupling is desirable. The high gain of the cascaded stages will, however, cause a wide output voltage swing with temperature by multiplying any small input change by the total DC gain of the amplifier, effectively shifting the baseline with respect to a fixed threshold. As would be expected, this causes changes in the false alarm rate and probability-of-detection of the system. If AC coupling is used, the DC drop across the coupling capacitor varies with the duty cycle of the environment, again shifting the threshold-to-base line settings, creating recovery time problems.

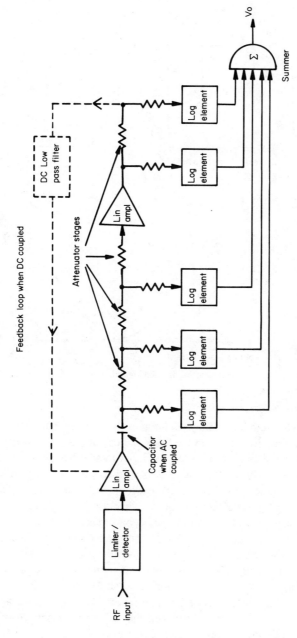

Figure 10-1. Detector logarithmic video amplifier.

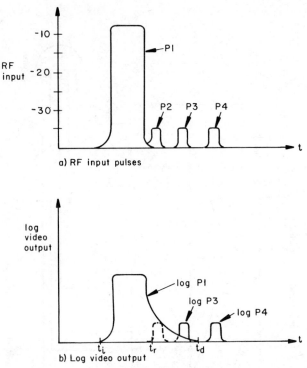

Figure 10-2. Recovery time problem in an AC coupled logarithmic video amplifier. (*a*) RF input pulses. (*b*) Log video output.

To achieve DC stability, many techniques have been attempted. In Figure 10-1, it may be seen that DC loop feedback is possible. Any DC drift can be fed back inversely from the output stage to the input to stabilize the circuit; however, careful attention must be paid to the low-pass filtering since any in-phase high-frequency component existing in the feedback loop will cause oscillation. Since the loop will enclose all ground returns, DC feedback logarithmic amplifiers tend to be unstable and sensitive to ground loop currents, making layout and shielding as much of an art as a science. Successful designs of DC loop amplifiers have been used for years, however, and this method is useful for moderate video bandwidth (B_v) devices. Oven stabilization is also used to reduce early stage drift in some designs. The other solution, AC coupling, also shown in the figure, is useful as long as very dense environments or CW signals are not encountered since the DC buildup across the capacitor will vary with the DC level associated with the first component of the Fourier series of each pulse train. Also, the DC recovered by detecting a CW signal is lost unless chopped or otherwise converted to an AC signal that lies within the video bandpass of the balance of the logarithmic amplifier.

An AC coupled logarithmic amplifier is prone to recovery-time problems. Since this is a major consideration, it must be carefully considered at the outset. Figure 10-2 details the effect. Assume in Figure 10-2a that four RF pulses are detected by the receiver. Pulse P_1 is at approximately -7 dBm, pulse P_2 to P_4 are at -35 dBm or 28 dB lower. The recovered video of P_1 is stretched at low video levels due to the shift and duty-cycle DC components of the environment as shown in Figure 10-2b. The result, therefore, is complete masking of P_2 and distortion of P_3. It is only after t_d that P_4 can be received and reliably processed. This effect increases the shadow time of the receiver by $t_d - t_2$, where $t_2 - t_1$ is the pulse width of the output video and t_d is the delay time. It is possible to have delays as great as or greater than the pulse width of the strongest pulse, desensitizing the receiver for extended periods. Solutions for this problem are to increase the AC coupling by using large capacitors, to DC couple, or to utilize a pulse-on-pulse type of DLVA, which is described below.

DLVA design starts with the intent to provide a saturating or cutoff plateau at various signal levels as determined by the attenuation and logarithmic elements. Hughes describes the design process well in an early paper (1) and in his book (2). Conceptually, each logarithmic video element saturates at a different signal level as determined by the gain of the linear amplifiers and attenuators. After saturation, the stage holds at a fixed voltage, which is added to the next stage. The unsaturated gain of each increment follows a logarithmic characteristic due to the input/output transfer characteristics of the device.

Figure 10-3 shows two typical logarithmic elements that depend upon the $V-I$ characteristics of a diode or diode-connection of a transistor. Temperature drift of this type of DLVA depends upon the saturation current variation with temperature as it passes through the bulk or spreading resistance r_b of the diode or transistor.

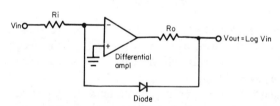

a) Feedback logarithimic element using diode V-I characteristics

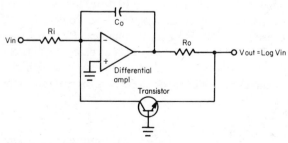

b) Feedback logarithimic element using transistor

Figure 10-3. Logarithmic element devices. (*a*) Feedback logarithmic element using *V-I* characteristics. (*b*) Feedback logarithmic element using transistor.

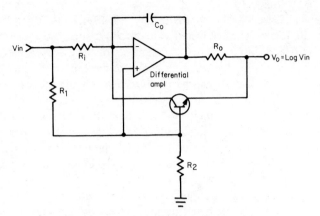

Figure 10-4. Stabilized logarithmic element stage.

Since this current approximately doubles for every 10°C temperature change, the DC stability for each stage changes, with the earliest stages multiplied by the gain of all successive stages, causing a large output variation. Figure 10-4 permits some improvement by returning the positive side of the differential amplifier to the voltage divider composed by R_1 and R_2. Reference 3 shows that if R_1 and R_2 are chosen such that

$$\frac{R_2}{R_1 + R_2} = \frac{r_b}{R_i}$$

the error term is canceled almost completely. The reference also shows a dual logarithmic amplifier that uses a matched pair differential feedback element to effectively stabilize both channels, which are then subtracted to form the logarithmic ratio in a dual-channel system.

The video logarithmic amplifier must be designed to approximate a straight line of the plot of E_o versus RF input to the detector as expressed in dBm. This is shown in Figure 10-5, which is an idealized plot of a compensated amplifier. In this case, more gain is added for high RF inputs (> -10 dBm) to make the output compensate for the diode detection departure from square law. Points P_1 to P_3 are the "break points," where each logarithmic element saturates. P_4 is where the stage that is active for large RF signals (> -10 dBm) adds its gain, which is purposefully made greater than the other stages to compensate for the detection law change of the detector. The compression ratio m, or slope of a logarithmic amplifier, is defined as follows:

$$m = \frac{\text{output dynamic range}}{\text{input dynamic range}} \quad \text{(in millivolts per decibel)}$$

Values of 27 mV/dBm are typical.

In a more recent design approach, Potson and Hughes describe a DC-coupled fast logarithmic amplifier capable of processing 10 nanosecond pulses. The approach is described in Refs. 4 and 5. An operational amplifier having an inverting and noninverting input is connected in a current feedback arrangement that forces both the noninverting and inverting input voltages to be equal, thus causing the inverting input impedance to be low. The operational amplifier is designed to have bandwidth rise and fall times independent of the closed loop gain. Figure 10-6 shows the configuration. Amplifier A_1, which operates over the entire dynamic range, drives operational amplifier A_2, which has a low input impedance as the result of the current feedback. The gain of A_1 is equal to the feedback resistance R_{fb} divided by the parallel equivalent resistance due to the parallel combination of the input resistors

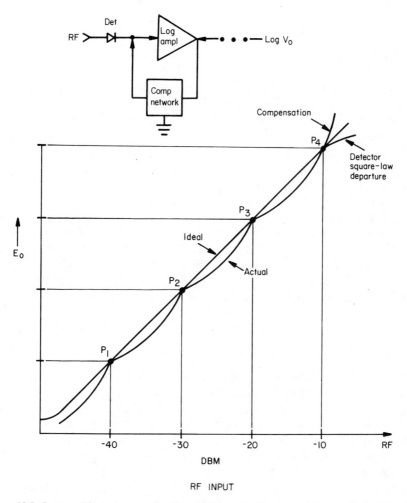

Figure 10-5. Log amplifier output as a function of RF input in dBm into a detector with compensation.

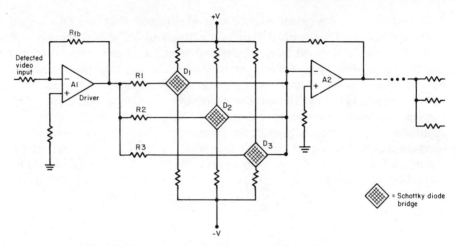

Figure 10-6. Current summed series sequential logarithmic amplifier.

R_1, R_2, and R_3 and each biased Schottky diode bridge D_1, D_2, and D_3. The diode bridge, which is the nonlinear logarithmic element, is biased and supplied with signals at various break-point levels as established by R_1, R_2, and R_3 acting as attenuators in conjunction with the dynamic bridge impedances. As the signal increases, the current to the operational amplifier A_2 is sequentially limited, increasing the parallel input resistance, which inversely reduces the loop gain in a logarithmic manner. In this way, many break points can be included in a single stage, reducing the number of stages and the problems associated with DC stabilization. Since the diode bridge is fast, the operational amplifier maintains its bandwidth as the gain varies, and many break-points can be used. The series-sequential design offers the possibility for excellent performance for narrow pulses. Since DF technology is moving up into frequency ranges where 10–50 nanosecond pulsewidth times may be expected, this design offers considerable promise for the future. Although DC stabilization may still be needed, the fewer number of stages required simplifies this task.

10.1.2   The Pulse-on-Pulse DLVA

The need to reduce the time wasted in the recovery process for strong signals has prompted development by the AEL Corporation and others of a daily-line "pulse-on-pulse" compressive logarithmic video amplifier. This amplifier, which is a variant of the DLVA, allows only a sample of the input pulse width to be amplitude-logged, thus allowing the amplifier to be ready to accept a second pulse immediately after the first pulse, even if it occurs during the width or "backporch" of the first. The technique makes use of a reflective delay-line to subtract the pulse from itself thus forcing the amplifier to exhibit base-line gain during the time after a portion of the pulse has been amplified, as shown in Figure 10-7. This permits reception of a second pulse before the first has disappeared. Detected positive-going pulses are amplified and fed to a delay-line, which is reactively terminated such that it reflects the input pulse, after delaying it by t_D and returning it to the input after another delay of t_D. By proper choice of constants, the delayed reflected pulse completely cancels all but the portion of the input pulse that was able to pass to node A (twice t_D) while the delay-line was being loaded. The output pulse is normalized in width to $2 \times t_D$ as a result.

The logarithmic amplifier consists of DC coupled gain stages A_2, A_3, and A_4. Each stage contains limiting diodes that only conduct at certain positive pulse swings; upon conduction, each diode adds its voltage through the N_1 through N_3 network, all of which is summed in R_L. The noise-limiter is used to suppress negative-going pulses and noise that are not used in subsequent processing. Additional compression can be provided by transistor logarithmic elements in the N networks. This reduces the output dynamic range more than a standard logarithmic amplifier. The curve of output voltage versus power input in dBm is a straight line when plotted on log–log graph paper (logarithmic amplifiers use linear–logarithmic plots for straight line outputs). The mathematical effect of this additional compression is to change the base of the logarithmic system. The pulse-on-pulse technique is fully applicable to the more familiar logarithmic amplifier.

Figure 10-8 shows the result of the pulse-on-pulse design. Two input pulses, P_1 and P_2, the latter occurring during the P_1 interval, are applied to the system. Only a pulse of a width from the leading edge of P_1 to twice the delay time propagates through the amplifier. The second pulse P_2 can therefore be detected and used even though it is actually present simultaneously with P_1. This dramatically reduces the shadow time of most systems, which, under ideal conditions, would be at least the width of P_1. Contrast this to the pulse recovery time problem shown in Figure 10-2. The output width of the pulses is not equal to that of the original signals, and pulse width information is therefore lost. DC coupling is applied around the entire amplifier for stability. Detector and the first-stage amplifiers are susceptible to noise pickup; consequently, it is desirable to mount the detector as close to the antennas

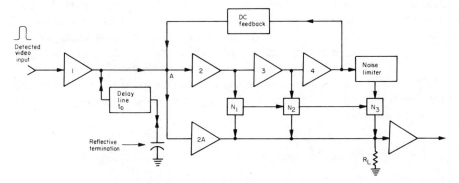

Figure 10-7. Pulse-on-pulse logarithmic compression amplifier.

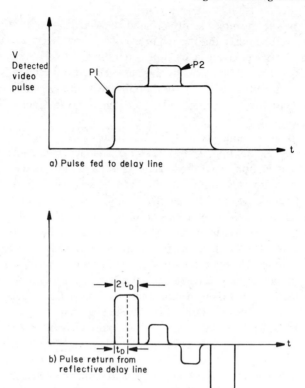

Figure 10-8. Pulse-on-pulse logarithmic amplifier operation on two closely spaced pulses. (*a*) Pulse fed to delay line. (*b*) Pulse return from reflective delay line.

as possible. This is also done to reduce RF losses in the connector cable. DLVA units used in radar warning receivers are therefore often placed in aircraft wingtips and tail sections, leading to the need for hybrid miniature designs to reduce size.

Log and compressive amplifiers may be built with discrete components or in hybrid form. A hybrid assembly consists of a miniature ceramic printed circuit, which has chip capacitors, resistors, diodes, and semiconductor devices inserted into mounting holes and bonded to it. This creates a subminiature package that is a hybrid combination of miniaturized and discrete components, which can be hermetically sealed itself and treated as a subcomponent. Figure 10-9 is a photograph of a production version of a hybrid DLVA manufactured by AEL in large quantities. Circuits of this type will undergo further packaging variations as space continues to be at a premium.

10.1.3 *Logarithmic RF Amplifier*

Over the years there have been many attempts to develop a true or "detectorless" RF logarithmic amp. This was the intention in the concept of the IAGC systems, discussed in Chapter 2, but physical settling time, regenerative effects, and un-availability of wideband amplifiers prevented the early development from proceeding. A true logarithmic RF amplifier offers the advantage of preservation of both phase and amplitude information, permitting many different types of combinational monopulse processors to be used. We have seen how both amplitude and phase detection can be used to resolve ambiguities in many systems and can be used in scanning systems for centering of the tuning local oscillators.

The true logarithmic RF amplifier that has been most successful was designed by Barber and Brown (Ref. 6) and is available as a product from Plessey Semi-conductors. The concept used was to cascade a series of dual-gain stages each consisting of an amplifier connected in parallel with a limiter as shown in Figure 10-10. Each duo pair consisted of an amplifier that linearly provided gain until the limiter stage limited, after which the output of the duo pair stayed constant. This constant output was fed to the next cascaded pair and formed a pedestal upon which the next pair built its gain until saturation, and so forth. This is shown in Figure 10.10a and b, where each breakpoint indicates where the stage saturation takes place.

Figure 10.10c shows how each duo adds up to form a logarithmic curve with a slope of

$$\frac{V_L}{20 \log(G + 1)}$$

where

$$V_L = \text{the limiter voltage}$$

$$G_1 = \text{the stage gain}$$

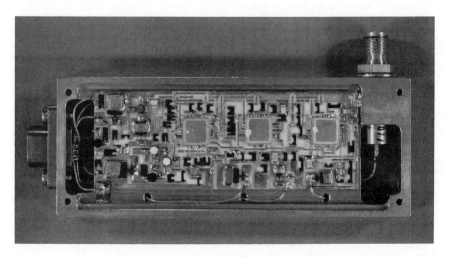

Figure 10-9. Hybrid DLVA. Courtesy of AEL.

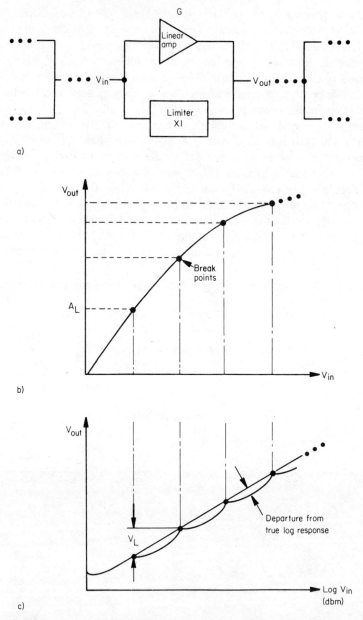

Figure 10-10. (*a*) Cascade of duo stage pairs consisting of an amplifier and limiter. (*b*) Resulting gain piecewise linear approximation. (*c*) Log characteristics.

Since there is basically no continuous logarithmic element, as is found in the diode or transistor case, the error or departure from the logarithmic response curve depends upon using as little linear gain as possible and therefore requiring many stages, which is not a particularly economical approach. Using hybrid packaging techniques, however, eight-stage true logarithmic amplifiers can be built with an 80 dB RF dynamic range compressed to a 10 V output range useful to 70 MHz. Phase variation, which was ±4 degrees in the early model, has been subsequently improved. Reference 7 presents an interesting curve that shows single stages versus peak logarithmic error in decibels for this design. A 12 dB gain stage will give 1 dB of error. An eight-stage design using 10 dB gain stages should attain a linearity of ±1 dB. The true RF logarithmic amplifier finds application mainly as a wideband IF amplifier in superheterodyne receivers, since it does not have a wide instantaneous RF bandwidth (B_r) as does the DLVA. It also finds use in radar receivers that require phase information moving target indicators, anticlutter, etc.) as well as amplitude data.

10.1.4 Successive Detection Logarithmic Amplifier

The successive detection logarithmic amplifier is, in actuality, a combination of the previous two designs. It consists of an amplifier chain of a cascade of linear amplifiers of n stages that provide maximum RF gain for weak signals and minimum gain for strong signals as shown in Figure 10-11. This amplifier has been described previously in the study of logarithmic IF systems, where it is most often used. Each stage of linear gain is detected and clamped to form a pedestal representing the maximum signal at that point. Each pedestal is in turn summed through a delay line to provide a logarithmic video output. The delay line is used to compensate for stage delay and to reduce recovery time. The RF output is not linear since the detection process distorts the total dynamic range; however, with careful design some gross phase information can be obtained.

There are several variations of the circuit shown. For IF applications, the stages may be coupled together by resonant interstage networks, the purpose being to limit the RF instantaneous bandwidth desirable for superheterodyne image rejection, and so on. In sum and difference applications, this may create a problem since the number of poles that determines the RF band pass becomes proportional to the signal strength. Since the sum channel will be of greater amplitude than the difference channel in null systems, the phase error due to dynamic range may be significant since the number of poles in use in each channel differs affecting phase. Logarithmic amplifiers compress the response of filters. Therefore, all filtering should be done with a multipole sharp slope filter (near ideal).

The successive detection amplifier as shown can be used for bandwidths up to those frequencies where parasitic effects in the detector diodes become a problem. It is possible to design the system with an iterative impedance of 50 ohms and follow microwave techniques; however, the DLVA can offer a less costly solution. For these reasons, the successive detection amplifier is used for video and IF applications in frequency ranges up to approximately 200 MHz. Care must be exercised in layout, however, to prevent instability due to high gain, and so on.

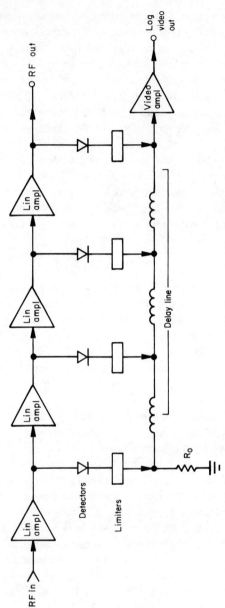

Figure 10-11. Successive detection logarithmic amplifier.

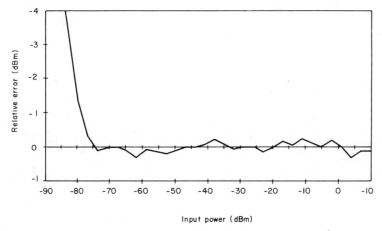

Figure 10-12. Matching characteristics of two 160 MHz successive detection logarithmic IF amplifiers. Courtesy of the RHG Corp.

There are commercial logarithmic RF amplifiers available capable of being matched to within ±.5 dB. Figure 10-12 is a plot of the departure from ideal logarithmic characteristics for an RHG corporation 160 MHz center frequency design. The amplifiers have a 30 MHz bandwidth response and exhibit the indicated match over an 80 dB dynamic range (with a slope of 25 mV/dB).

10.2   A PHASE CORRELATOR FOR WIDEBAND APPLICATIONS

The wide RF bandwidth phase correlator is, in a way, analogous to the wide bandwidth detector logarithmic amplifier since it provides a means to measure differences in phase to a high degree of precision. The first use of the phase correlator was to measure the differences in phase resulting from the time delay in one path of a dual output coupler. The insert in Figure 10-13 shows the idea: Any signal fed to the RF input would arrive as two voltages, where the differences between the angles A and B were proportional to the time delay t established by the delay line. The purpose of the correlator, shown in the remainder of the figure, is to provide sine and cosine voltages of the angle ϕ, which are proportional to and a measure of the instantaneous frequency of arrival. A polar display with radius (ρ) was proportional to signal amplitude; the angle (θ) was proportional to frequency. Later implementations made use of a network of various frequency encoders, each offering different delay times to provide ambiguity resolution over a wide frequency range. The concept has been well described in the literature from early reports to more current information (8–10). The significant result of the development of this technology has been the design of practical phase correlators and digitizers capable of being used for monopulse DF processing. The technique was outlined in Chapter 5, Section 5.4, in the description of the moded/Butler circular array.

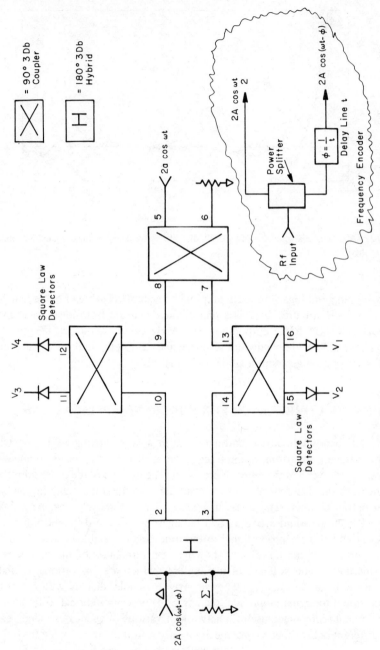

Figure 10-13. Phase correlator, frequency encoder shown in insert.

278

The basic inputs to the phase correlator are two amplitude limited voltages that differ in phase by ϕ, which is the angle containing the desired DF information. In Figure 10-13, the input voltage at port 5 of 90 degree hybrid is $2A \cos \omega t$, whereas the other input containing the phase information enters the 180 degree hybrid at port (1) (Δ) and is of the form $2A \cos (\omega t - \phi)$. The action of each coupler is traced and is listed below by port number for identification:

Ports	Voltage
2–10	$\sqrt{2A} \cos \omega t$
3–14	$-\sqrt{2A} \cos \omega t$
8–9	$\sqrt{2A} \sin (\omega t - \phi)$
7–13	$\sqrt{2A} \cos (\omega t - \phi)$

These are the internal voltages. The output voltages summed through the hybrids appear at the ports, just ahead of the detectors, as follows:

Ports	Voltage
16	$A \cos (\omega t - \phi) - A \sin \omega t$
15	$A \sin (\omega t - \phi) - A \cos \omega t$
12	$A \sin (\omega t - \phi) + A \sin \omega t$
11	$-A \cos (\omega t - \phi) + A \cos \omega t$

If a square-law detector is placed at the output of each of the above ports, detection takes place in two stages; first the above voltages are squared:

Port	Voltage
16	$A^2 \cos^2 (\omega t - \phi) - 2A^2 \cos (\omega t - \phi) \sin \omega t + A^2 \sin^2 \phi$
15	$A^2 \cos^2 (\omega t - \phi) - 2A^2 \sin (\omega t - \phi) \cos \omega t + A^2 \cos^2 \phi$
12	$A^2 \sin^2 (\omega t - \phi) - 2A^2 \sin (\omega t - \phi) \sin \omega t + A^2 \sin^2 \phi$
11	$A^2 \cos^2 (\omega t - \phi) - 2A^2 \cos (\omega t - \phi) \cos \omega t + A^2 \cos^2 \phi$

then the detection process removes (filters) all of the RF (ωt) components leaving

$$V_1 = A^2 (1 - \sin \phi)$$

$$V_2 = A^2 (1 + \sin \phi)$$

$$V_3 = A^2 (1 + \cos \phi)$$

$$V_4 = A^2 (1 - \cos \phi)$$

The outputs of the detectors are usually applied to four differential amplifiers, which provide the voltages:

$$V_2 - V_1 = 2A^2 \sin \phi$$

$$V_3 - V_4 = 2A^2 \cos \phi$$

The cosine channel is called the in-phase or I channel; the sine channel, the quadrature or Q channel. The ratio of I/Q will give the tangent of ϕ, and the differential outputs can be connected to a polar display to give the circular $\rho-\theta$ display discussed above. The correlator for the purposes of passive direction-finding performs an analog-to-digital conversion, converting the two voltages into a digital word for the angle ϕ. This is readily done by synchro-to-digital counters and zero-crossing detectors and is a standard digital technique.

The accuracy of the simple encoder shown here is limited by the signal-to-noise ratio of the signal presented to the detectors and the encoding accuracy, which is usually only 5 bits. In the frequency encoder, the correlator shown would only represent one set of digital numbers, the most significant bits, and would only be encoded to $2^5 = 32$ places. A second discriminator, utilizing another longer delay line, would have one ambiguity but would have a greater slope of degrees/MHz. This second channel would provide the next group of lesser significant bits, and the process would continue with the addition of more encoders. In actual practice, an IFM usually has four delay line correlators of t, $4t$, $16t$, and $64t$, and instantaneous frequency measurement receivers can provide a 12-bit digital output. Systems of this type are called "gas-meter" encoders, in recognition of their similarity.

DF phase measurement utilizing multiple-fed arrays usually provides multiple phase angle ϕ outputs. This was shown for the circular array in Chapter 5, Section 5.4, by moding the Butler-fed array so that ϕ, $2 \times \phi$, $4 \times \phi$, and $8 \times \phi$ were available. The same technique is used in interferometers, as described in Chapter 6. In this latter case, circuit phase comparisons are made to resolve the N ambiguities. Phase encoders are generally preceded by an RF amplifier and hard (sharp knee) limiter to remove all amplitude variation, since the detector outputs are proportional to both the instantaneous amplitude and phase differences in the I and Q channels. The limiting action provides an amplitude-stripping action. Since limiting creates harmonics, any output harmonic power is subtracted from the fundamental. A signal slightly weaker than the desired signal will therefore be rejected, the phenomenon being the well-known "capture effect."

Phase discriminators do suffer from accuracy problems in the presence of two nearly equal signals, whether pulse or CW. If the input signals differ by more than 6 dB, the effects are minimal. The presence of jamming noise or a CW signal, however, can create inaccuracies and desensitization of the receiver. For these reasons, phase discriminators are often preceded by tunable filters that can pass or reject bands of frequencies. The difficulties in instantaneous frequency measurement using phase discriminators is gradually leading to the use of DF driven receivers and to the development of true digital instantaneous frequency encoders.

10.3 DF PROCESSING

Microwave passive DF provides the means to determine the angle-of-arrival of a signal from either a single pulse or from many pulses. In the first case, high signal-to-noise penalty must be paid; in the latter case, high precision accuracy can be achieved. The immediacy of the need for the information in part determines the system configuration. Processing of the DF intercept descriptor falls into two categories as well. First, there is the preprocessing that identifies that the signal is present and at a sufficient level to fall within accuracy, false alarm rate, and probability of detection criteria; that is, are the data usable? The second category uses the information in various ways. ESM and signal intercept systems in the early 1970s used time data such as pulse time-of-arrival and duration, amplitude, and DF as sorting parameters, since it was easy to determine the nature of a threat by the pulse groupings, signal levels, and type of emitter scan. Frequency was used to identify the band of operation, allowing the combination of parameters to be tested against a priori data stored in a computer memory to provide threat identification. It was simple to identify threats in a non-real-time environment since the design of threat radars were sufficiently different from friendly radars in pulse, scan, and modulation characteristics to permit fairly accurate characterizations. Unfortunately, as stated previously, the situation has changed; enemies became friends, friends became enemies, and equipment from all sources began to become dispersed creating homogeneous threat environments difficult to reduce to simple dimensions. The first answer was the addition of the IFM as an additional sorting word, and many processors were designed to add pulse-by-pulse frequency as a signal tagging or correlating word. ESM equipment using this method is categorized as "frequency driven" as stated many times throughout this book. The limited IFM encoder–correlator approach in many different configurations was perfected and available for both base band (e.g., 2–6 GHz) and direct band (e.g., 2–8, 8–18 GHz) applications. This solution has carried ESM technology into the 1980s; however, new factors have elevated the DF descriptor into passive consideration.

Instantaneous frequency systems do not operate well in environments where there are many signals at amplitudes similar to the threats being detected. There should be at least a 3–6 dB difference in amplitude for accurate measurements to be made. In a light environment, the possibility of pulse coincidence is less than in a heavy environment; but as time progresses, the environmental pulse densities seem to increase. The presence of CW or high pulse repetition frequency (PRF) signals further exacerbate the problem. Although there are methods to recognize and separate CW signals (11), and filters, such as using tunable YIGs to accept or reject desired and undesired signals, frequency has diminished somewhat as the key sorting parameter. This trend has been accelerated by the proliferation of frequency agile and frequency diversive radar systems.

Direction-of-arrival, in contrast, has become more significant, with true monopulse bearing–taking becoming an important factor because the DF word is less difficult to measure with the availability of faster digital hardware and because DF cannot rapidly change. A threat can vary frequently very rapidly or use frequencies other than usual to confuse frequency-driven receivers. Angle-of-arrival of a signal is hard to change quickly since it depends upon physical factors of the ballistics of a missile or a plane. It is admittedly hard to change DF quickly for ground targets.

Monopulse radar technology is another factor operating to enhance the use of passive direction finding to sort signals. As monopulse radars become deployed more extensively, the use of scan modulation disappears (monopulse radars are constant illuminators), and intercept signal scan cannot be used to sort signals. The ability to jam monopulse radars requires rapid response, and often DF data are the only certain parameter available to "set-on" electronic countermeasures or to direct chaff. In a contrasting sense, the improvement in RF sensitivities and the ability to package greater computer aspects into warning or ELINT systems permits monopulse to be used more accurately for angle measurement.

A good example of DF technology is in the use of the interferometer receiver. It is clear that this type of system is capable of high resolution at the price of narrow bandwidths or of multiple ambiguities for wide-band applications. With the ability to add computer assets and to combine amplitude and phase measurement techniques, these ambiguities can be resolved. Practical interferometer DF receivers are now deployed successfully. Many years have passed since the early interferometers were developed, set aside, and redesigned using modern techniques. Much of the popularity of the interferometer is in its ability to obtain accurate DF measurements.

10.3.1   Non-Real-Time DF Processing

DF determination processing is the implementation of the unambiguous measurement of the angle-of-arrival of a threat. This can be accomplished by comparison of amplitudes (or phases) of adjacent channels if the signals in the two channels to be compared are at the proper signal-to-noise ratio. What should be done if they are not? The answer is not always apparent; some systems reject both the strongest and next strongest signals and wait for the next acceptable pulse pair. Other systems indicate boresight reception in the direction of the strongest of two pulses with or without a flag to calibrate the measurement credibility. Some systems average signals and merely start a "track" or file of the possibility of a signal in a wide sector "bin," with hopes of reducing the bin size as more signals appear. Other more adaptive systems look for other descriptors such as frequency or wait or retune to optimize the situation.

A DF receiver may be reactive or adaptive depending upon the design philosophy. The actual DF determination process usually starts with a recognition of the most significant bit of the DF measurement, which can be simply recognition and codification of the antenna that picked up the one pulse. The balance of the DF determination will make use of the direction-finding resolution capabilities of the receiver. In the case of a rotating DF antenna, the receive bandwidth and pointing angle can be encoded; in monopulse systems, the ratio can be formed using amplitude and/or phase processor technology. In most cases, however, the DF word is formed in real time and as quickly as possible. This is almost an absolute requirement for DF driven systems and is part of the reason for the development of the pulse-on-pulse techniques described above.

Figure 10-14 is another view of a generic receiver that operates in non-real time; that is, the output data is processed independently of the input data and is not time synchronized with it. This type of system takes a picture in time of the signal environment and loads it into one-half of a temporary storage (usually a random access memory, or memory (RAM). Measurements are supplied to a threat memory as a set of fields for DF, amplitude, time of arrival, and frequency either as an RF band (from RF multiplexed systems) or as a word (from an IFM). The suitability of a signal is then determined by a set of threshold detectors operating on the n (in this case four) DF outputs to indicate that the signal-to-noise thresholds have been crossed and that a signal is suitable. Signals so judged are fed to the monopulse ratioformer and the omni video and RF band analog-to-digital converters and to the pulse wordformer where all immediately available information is grouped. Each pulse group or set of fields is loaded into the active threat memory half, whereas the other half is operated on by the central processing unit (CPU). A nonvolatile program memory holds the algorithms for the CPU and, in the case of a warning receiver, may be loaded prior to a mission. In the case shown here, two memories are used: One contains the program, the other the threat a priori database or "telephone directory." The CPU compares the static world of pulse fields until periodicity or match is obtained. The pulses are separated in non-real time and the parameters are compared to the a priori data for identification. If ambiguities arise, the processor may call for IFM frequency data, if only bands were used, or may classify the threats as unknown. The threat priority file keeps track of each threat and prioritizes them by lethality and range for display.

As described above, the non-real-time system works well in moderate environments. DF determination is done in conjunction with the monopulse time available data and can be used for sorting by using only the DF field as the sorting tool. When environments are dense and the switched memory system is unable to operate, DF inputs can be used to limit the sector of view to quadrants or octants to "wane" or thin the environment to permit the system to shed some of the work load. Access to the DF system is usually available at the control indicator where limits can be manually set. To do this the DF output word must be readily available. DF data when stored in the memory for computation and presentation must be in "true," or north reference, form since the host vehicle is usually constant motion, consequently compass heading information is required. Since most monopulse ratioformers are analog (subtractors, discriminators, etc.), an analog-to-digital conversion is often made. Since intercept pulses of about 100 nanoseconds are expected, it is usually desirable to normalize all pulses to some wider standard width to permit a common A-to-D conversion technique to be used. This can be done by the subtractor used with the DLVA or by the pulse-on-pulse DLVA by choice of the delay-line as explained above. The phase correlation system provides a digital word directly since in most phase monopulse systems, the development of the ratio is done in the encoding process. Frequency, in the form of a digital word, is requested by the

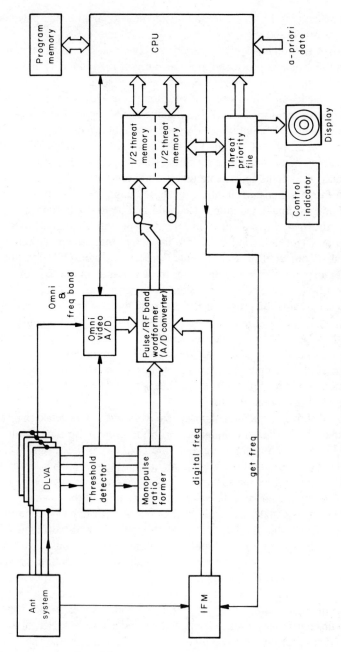

Figure 10-14. DF process functions in a non-real-time generic receiver.

computer to resolve ambiguities, and so on. IFMs take time (300 nanoseconds typically) and therefore are generally used by request in the DF driven system shown here.

10.3.2 The Real-Time DF Sort Preprocessor

The DF word may be used as the first filter in the receiver as real-time sorting parameter. To do this, our generic receiver is divided into preprocessor and processor sections. This restricts the information flow to the non-real-time processor on the basis of real-time preprocessor filtering. Figure 10-15 shows the preprocessor concept as used to filter data flow to the computer. In this case, the subtle difference is that *every* pulse is DF tested against limits *before* the data are passed to the computer. This reduces the computer loading. The monopulse pair is formed by a high-speed scan of the outputs of the DF antennas, or of a DF and omnidirectional antenna pair, using DLVAs and two parallel receiver channels. A single channel supercommutated around the *n* DF antennas within the width of a single pulse or a parallel channel DF system with an analog sample-and-hold encoder can be used to provide real-time preprocessor DF data.

Since the DF sort preprocessor is designed to use the DF word as a decision maker or to group or associate pulses by DF to reduce the computer work load, consistent DF measurements must be made on all signals. This requires high DF accuracies, and consequent good signal-to-noise ratios for each pulse which reduces sensitivity. As a further aid, it is not unusual to add digital correction to each angle on the basis of azimuth or frequency. Precalibration of the host vehicle is another means to generate a usable azimuth-of-arrival correction data base.

10.3.3 Database Generation

Both of the above generic systems presuppose that a convenient telephone directory of possible threats can be entered into the system for sorting intercepts. How is this table directly developed? The answer is by electronic intelligence (ELINT) systems. The systems used for this purpose are large and usually unprogrammed at the start.

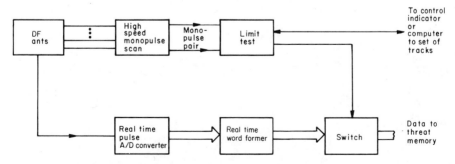

Figure 10-15. Preprocessor for real-time DF processing.

The receivers are used with no a priori filtering to build up a database that is examined for possible groupings of likely threats. Synergistic information from other sources such as visual, photographic, and infrared is used to identify locations and directions; association between the ELINT electronic data and the other sources is obtained, and appropriate threat tables are prepared. In this way, the a priori data can be developed and updated. For this reason continual ELINT survey activity is required to allow new corrections to existing data to be made to filter threats as well as friendly signals as they are so identified.

10.4   DF DISPLAYS

The output of a completely digital direction finding system can be used to set electronic countermeasures on the target automatically, or the output data can be recorded for postmission analysis. In either case, a real-time display or record of activity is ultimately required. There are many ways to group displays. The simplest type is a rho (ρ), theta (θ) analog display of bearing versus-signal-strength as discussed in Chapter 2, Section 2.1, and illustrated in Figure 10-16. In this type of presentation, the actual video of the monopulse signals is stretched in amplitude and presented as a "strobe line" on a CRT. The earliest analog types of radar warning receivers used this technique to indicate threat presence. A 3½-inch diameter, high-intensity CRT display indicated the presence of threats by integrating pulses using the tube phosphor. Larger types of CRT displays of this type were and are presently used for ELINT DF receivers. In these simple displays, the operator is alerted to the fact that there are threats around him. In airborne warning applications, the bearings of the threats are relative and identification can be accommodated by selecting the frequency threat bands to be displayed, assuming that certain types of threats, such as airborne intercept or ground-to-air types, are in the displayed bands.

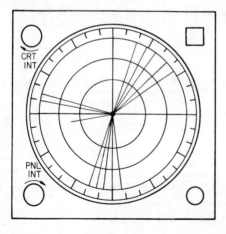

Figure 10-16. Analog RWR strobe display.

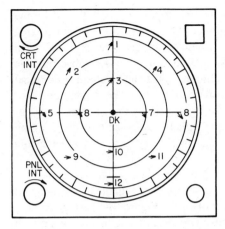

Figure 10-17. Alphanumeric radar warning display.

Needless to say, for many applications this early method provided an unsatisfactory presentation. A significant problem was that fluctuation in signal strength caused serious variation in the angle of the threat that was presented. As threats moved in and out of range, the strobes occupied a large sector of the display, and visual bisecting of an area by the operator was necessary. Some assistance was provided by allowing the operator to listen to the video of the actual intercepted radar pulses. These short pulses were stretched in width to be made audible. Preservation of the relative amplitude of the audio permitted the scan characteristics of the signals to be heard, and since certain scanning tones could be associated with certain threats, the operators could recognize a missile launch or target illumination signal and take defensive action. The equipment described was used in the late 1960s in the Vietnam War era. It became obvious that as a warning display, the simple analog polar type was inadequate.

DF displays for ELINT and large systems, such as found in land-based or shipboard applications, used DF displays that were polar or rectilinear. These displays plotted DF data as a plus or minus variation from a known angular reference. In these applications, much more data were available; parametric pulse information was presented on five-gun multiple scan CRTs using different scans for each trace to permit pulse width, PRF, and scan characteristics to be obtained. To this, the frequency to which the receiver was tuned could be added to complete the set of intercept descriptors. It is noteworthy that one shipborne piece of equipment, the AN/WLR1, using displays of this sort, has been and is presently in operation after almost 30 years. As threats proliferated, digital data processors were added and the output displays changed. Threat identification was now made by the methods described above, and the output displayed (and often simultaneously recorded) was in alphanumeric form.

Figure 10-17 is an airborne radar warning display showing how an alphanumeric presentation is used to present a large amount of information to the operator. In the display shown, the threat type is represented by a symbol that is placed at the angle-of-arrival. This angle can be either true or relative, as selected by the operator.

The lethality of the signal, represented by its signal strength, is indicated by the radial distance from the outer edge, the graticule circles representing increasing priority. Thus a threat in the center circle would be of the greatest concern. In addition, the status of the threats acquisition and control systems, as determinable from its modulation, is indicated by symbol blinking and an audio alarm. Additional threat-type information is shown by the arrow indicators. If the display becomes congested (many threats in one quadrant), the lesser priority threats are dropped or the display can be expanded to show multiple signals at a given crowded azimuth sector. This type of display is typically of high brightness to permit viewing in a sunlit aircraft cockpit.

ELINT displays also make use of digitally processed data. A typical display can be polar as discussed above; but since more screen space is available in a rectangular format, positional DF data display is not generally used. Figure 10-18 is a frequency-azimuth threat display found in large systems. This display is a frequency-azimuth presentation showing alphanumeric identified threats as they range from a bore-sight azimuth (along the abscissa) at the frequency of occurrence (on the ordinate). The identity of the threat is designated by the symbol denoting recognition of the class of threat represented. A cursor controlled by a joystick hooks any individual

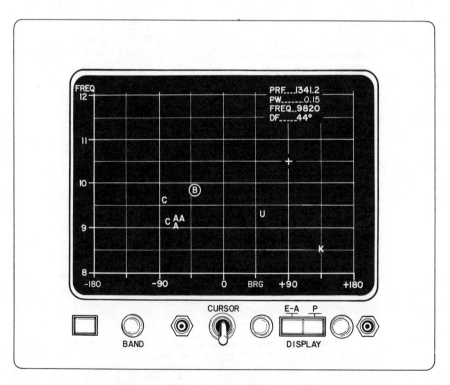

Figure 10-18. Frequency-azimuth threat display.

threat by encircling it as shown. A parametric readout of the encircled threat is displayed in the corner and is recorded together with header information consisting of date, time, mission number, vehicle location, and so on, for postmission analysis. The frequency-azimuth display finds favor in shipboard systems since the boresight can be made to represent the forward direction or locus of travel of the ship. This is a convenient way to identify angle quadrants quickly and permits evasive maneuvering or countermeasures to be used.

REFERENCES

1. Hughes, R. S., "Make Very Wide-Range Log Amps Easily," *Electronic Design*, Oct. 11, 1970, pp. 76 ff.

2. Hughes, R. S., *Logarithmic Video Amplifiers*, Dedham, MA: Artech House, 1971.

3. Morgan, D., "Get the Most Out of Log Amplifiers by Understanding Error Sources," *Electronic Design News*, Jan. 20, 1973, pp. 52 ff.

4. Potson, D., and R. Hughes, "Extremely Fast Log Amp Handles Narrow Pulses," *Microwaves & RF*, Apr. 1985, p. 85.

5. Potson, D., and R. Hughes, "DC-Coupled Video Log Amp Processes 10-ns Pulses," *Microwaves & RF*, May 1985, Part II, p. 75.

6. Barber, W., and E. Brown, "A True Logarithmic Amplifier for Radar IF Applications," *IEEE Journal of Solid State Circuits*, Vol. SC 15, No. 3, June 1980, p. 291.

7. Lansdowne, K., and D. Norton, "Log Amplifiers Solve Dynamic-Range and Rapid-Pulse Response Problems," *Microwave Systems News*, Oct. 1985, pp. 99 ff.

8. Wilkens, M. W. and W.. R. Kincheloe, Jr., "Microwave Realization of Broadband Phase and Frequency Discriminators," Stanford Electronics Labs, Technical Report 1962/1966-2-SU-SEL 68-057, Stanford, CA, 1968.

9. Lipsky, S. E., "Measure New Threat Frequencies Instantly," *Microwaves*, Dec. 1970.

10. Gysel, U., and J. P. Watjen, "Wide-Band Frequency Discriminator with High Linearity," *IEEE MTT-S Internal Microwave Symposium Digest*, 1977.

11. Tsui, J., and G. Schrick, "Instantaneous Frequency Measurement Receiver to Separate Pulse and CW Signals," U.S. Patent 4,1194,206, Mar. 18, 1980.

Chapter Eleven ————————————————————

Future Trends

Microwave passive direction finding has been presented here in many different forms and many different ways. As originally stated, the purpose of this book has been to trace both the development of direction finding from its origins and to explain and describe the means of the technology to permit interested engineers to find the methods of the technique in one place. Any book covering as many aspects as I have tried to do undoubtedly raises questions. Some can be answered by more research, some by intensive study of the references, some only by the future acting as a driving force.

The rapid development of solid-state devices capable of developing RF energy at high microwave frequencies has resulted in small, high-gain, active transmitting and RF receiving antenna array systems. These arrays will permit high-effective radiated power to be generated by multiple aperture antennas, which will require high-accuracy, steered-beam receivers for seton. Such systems presently exist and are becoming deployed in many new applications. Small size is a natural fallout, and it is reasonable to expect printed circuit RF techniques to become more prevalent for both the circuitry and radiating elements. Evidence of this trend is clear in present literature, with the strong emphasis on microstrip and "patch" antenna technology. Adaptive array systems are also being developed to optimize these concepts.

Passive DF accuracy has become an important targeting or aiming tool, forcing DF accuracy to be improved. These accuracy trends for future radar warning or ESM systems place emphasis on built-in antenna and receiver systems, using both amplitude and phase interferometer techniques to measure and sort signals on a true monopulse basis. Most new electronic support measures receivers include instantaneous frequency measurement of baseband downconverted signals to provide additional intercept sorting and recognition capability. The trend in processing has followed the concept of built-in calibration to achieve angle-of-arrival accuracy measurements and resolution to unheard of limits. Multiple baseline interferometers appear to be increasing in importance, working in combination with amplitude monopulse techniques to augment these objectives.

The ability to generate high power in millimeter wave frequency ranges (greater than 20 GHz) permits systems to achieve higher spatial range. This requirement, in turn, dictates the use of high pulse repetition frequencies and narrower pulses (25–50 nanosecond pulse widths are not uncommon). Both ELINT and warning receivers must be capable of covering millimeter frequencies with sufficient sensitivities to overcome the high atmospheric attenuations at these frequencies. New technologies abound in this area. There are antenna systems, such as the Antector, that can be used to replace existing 2–18 GHz antennas to provide both crystal video and superheterodyne sensitivities well into the millimeter wave range. The classic methods of Gaussian pattern monopulse antenna techniques are being extended with multibeam DF antennas and arrays.

Future trends in passive detection will make use of additional information contained in the passive DF target returns. In current methods only the simple time presence of the intercept is used to make the DF determination, using the methods outlined in this book. A further extension of the technology can be used to derive parametric data. This can be done, for example, by analysis of the phase, amplitude, and times-of-arrival of each of the multiple time returns occuring in a given frequency bandwidth or by analysis of the multiple frequency returns occuring in the width of a DF cell. It is also possible to determine the range of an intercept passively, by noting the rate-of-change of DF amplitude or phase. An introduction to these concepts may be found in reference 1. The concept of detailed parametric analysis of the DF return to obtain additional data can reveal more information about the nature of the target; for example, the size of a plane or the presence of multiple targets (e.g., a formation of aircraft).

Processing and display technology has also kept pace with the RF developments cited above. New memory techniques permit large a priori data bases and complex operating algorithms to be stored for more powerful computational power. Faster throughput and parallel imbeded microprocessor methods will assure that systems of the future will accommodate the rapidly expanding and changing environment. Displays will be planar, LED, or flat CRT types that are bright and highly operator interactive. Direction finding will also extend into the optical and infrared frequency bands utilizing appropriate sensors and, in many cases, operating methods of the lower microwave DF technology. Expansion below the microwave band, although not the topic of this book, will also be given new impetus by the need for communications intelligence (COMINT) and the necessity for long-range direction finding and emitter location.

Microwave Passive Direction Finding portends the future by describing the past and present. The pace of development in DF, radar, processing, and displays is rapid and technically exciting. One can only hope that this technology, as well as that of companion surveillance and signal determination and recognition methods, will be dedicated to defensive electronic warfare for the good of all. To this goal, this book has been dedicated.

REFERENCE

1. Johnson, D., "The Application of Spectral Estimation Methods to the Bearing Estimation Problem." *Proc. IEEE*, 1982, pp. 1018–1028.

Index